A d v a n c e s   i n

Geosciences

Volume 14: Solar Terrestrial (ST)

ADVANCES IN GEOSCIENCES

Editor-in-Chief: Wing-Huen Ip *(National Central University, Taiwan)*

A 5-Volume Set

Volume 1:   Solid Earth (SE)
                ISBN-10 981-256-985-5

Volume 2:   Solar Terrestrial (ST)
                ISBN-10 981-256-984-7

Volume 3:   Planetary Science (PS)
                ISBN-10 981-256-983-9

Volume 4:   Hydrological Science (HS)
                ISBN-10 981-256-982-0

Volume 5:   Oceans and Atmospheres (OA)
                ISBN-10 981-256-981-2

A 4-Volume Set

Volume 6:   Hydrological Science (HS)
                ISBN-13 978-981-270-985-1
                ISBN-10 981-270-985-1

Volume 7:   Planetary Science (PS)
                ISBN-13 978-981-270-986-8
                ISBN-10 981-270-986-X

Volume 8:   Solar Terrestrial (ST)
                ISBN-13 978-981-270-987-5
                ISBN-10 981-270-987-8

Volume 9:   Solid Earth (SE), Ocean Science (OS) & Atmospheric Science (AS)
                ISBN-13 978-981-270-988-2
                ISBN-10 981-270-988-6

A 6-Volume Set

Volume 10: Atmospheric Science (AS)
                ISBN-13 978-981-283-611-3
                ISBN-10 981-283-611-X

Volume 11: Hydrological Science (HS)
                ISBN-13 978-981-283-613-7
                ISBN-10 981-283-613-6

Volume 12: Ocean Science (OS)
                ISBN-13 978-981-283-615-1
                ISBN-10 981-283-615-2

Volume 13: Solid Earth (SE)
                ISBN-13 978-981-283-617-5
                ISBN-10 981-283-617-9

Volume 14: Solar Terrestrial (ST)
                ISBN-13 978-981-283-619-9
                ISBN-10 981-283-619-5

Volume 15: Planetary Science (PS)
                ISBN-13 978-981-283-621-2
                ISBN-10 981-283-621-7

Advances in

Geosciences

Volume 14: Solar Terrestrial (ST)

Editor-in-Chief

Wing-Huen Ip

National Central University, Taiwan

Volume Editor-in-Chief

Marc Duldig

Department of the Environment,
Water, Heritage and the Arts,
Australian Antarctic Division, Australia

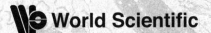

World Scientific

NEW JERSEY · LONDON · SINGAPORE · BEIJING · SHANGHAI · HONG KONG · TAIPEI · CHENNAI

Published by

World Scientific Publishing Co. Pte. Ltd.

5 Toh Tuck Link, Singapore 596224

USA office: 27 Warren Street, Suite 401-402, Hackensack, NJ 07601

UK office: 57 Shelton Street, Covent Garden, London WC2H 9HE

British Library Cataloguing-in-Publication Data
A catalogue record for this book is available from the British Library.

ADVANCES IN GEOSCIENCES
A 6-Volume Set
Volume 14: Solar Terrestrial (ST)

ISBN-13 978-981-283-610-6 (Set)
ISBN-10 981-283-610-1   (Set)
ISBN-13 978-981-283-619-9 (Vol. 14)
ISBN-10 981-283-619-5   (Vol. 14)

Typeset by Stallion Press
Email: enquiries@stallionpress.com

Printed by FuIsland Offset Printing (S) Pte Ltd. Singapore

EDITORS

Editor-in-Chief: Wing-Huen Ip

Volume 10: Atmospheric Science (AS)
Editor-in-Chief: Jai Ho Oh
Editor: Gyan Prakash Singh

Volume 11: Hydrological Science (HS)
Editor-in-Chief: Namsik Park
Editors: Joong Hoon Kim
                    Eiichi Nakakita
                    C. G. Cui
                    Taha Ouarda

Volume 12: Ocean Science (OS)
Editor-in-Chief: Jianping Gan
Editors: Minhan Dan
                    Vadlamani Murty

Volume 13: Solid Earth (SE)
Editor-in-Chief: Kenji Satake

Volume 14: Solar Terrestrial (ST)
Editor-in-Chief: Marc Duldig
Editors: P. K. Manoharan
                    Andrew W. Yau
                    Q.-G. Zong

Volume 15: Planetary Science (PS)
Editor-in-Chief: Anil Bhardwaj
Editors: Yasumasa Kasaba
                    Paul Hartogh
                    C. Y. Robert Wu
                    Kinoshita Daisuke
                    Takashi Ito

REVIEWERS

The Editors of Volume 14 would like to acknowledge the following referees who had helped review the papers published in this volume:

Nanan Balan
Hilary Cane
I.V. Chashei
Peter Chi
Yen-Hsyang Chu
Len Culhane
P. Demoulin
Marc Duldig
Malcolm Dunlop
Geza Erdos
R.A. Fallows
Brian Fraser
T.A. Fritz
Ivan Galkin
Chaosong Huang
John Humble
John Kennewell

Axel Korth
James LaBelle
Bo Li
Jun Lin
D. Maia
C. Mandrini
P.K. Manoharan
Ken McCracken
Karim Meziane
Y.-J. Moon
Jan Soucek
J.-N. Tu
Jingxiu Wang
Y.-F. Wang
Andrew Yau
Q.-G. Zong

CONTENTS

Advances in Geosciences
Vol. 14: Solar Terrestrial (2007)
Eds. Marc Duldig et al.
© World Scientific Publishing Company

ULF WAVES: EXPLORING THE EARTH'S MAGNETOSPHERE

B. J. FRASER

*Centre for Space Physics, School of Mathematical and Physical Sciences
University of Newcastle, Callaghan NSW 2308, Australia*

Ultra-low frequency (ULF) waves in the 1–100 mHz band are ubiquitous in the magnetosphere. They are a manifestation of hydromagnetic wave activity generated by physical processes resulting from the interaction of the solar wind with the Earth's magnetosphere. The almost pure sinusoidal signature of ULF waves when seen on the ground and in space by satellites suggests a resonance phenomenon. This is not unexpected since the size of the magnetospheric cavity is of the same order as the wavelength of the ULF waves. The magnetopause, plasmapause, and ionosphere provide convenient boundaries for wave reflection and transmission. Within the cavities bounded by these surfaces resonances may be established from propagating modes. These include the Alfven wave mode which is field aligned, resulting in field line resonance and the fast mode wave which propagates isotropically and may establish cavity or waveguide resonances in the magnetosphere and the ionosphere. It will be shown how the propagation and resonance characteristics of ULF waves can be used as diagnostic tools to determine important information on the topology of the dynamic magnetosphere and its plasma, from both spatial and temporal perspectives. These primarily employ simple and inexpensive ground-based instrumentation.

1. Introduction

Exploration of the near-Earth space environment, bounded by the extent of the geomagnetic field, has been primarily been undertaken using *in situ* Earth orbiting satellites. This commenced with the launch of Sputnik in the 1950s and also included the International Geophysical Year (IGY in 1957–1958). Through this international scientific collaboration we came to appreciate what could be discovered about the atmosphere which envelopes the Earth. Using very low frequency whistler observations, Storey[1] provided the first evidence that the region above the ionosphere at altitudes beyond 500 km was not just a vacuum. The similarity of VLF wave signatures in the two hemispheres indicated a conjugate association in a highly ionized region out to ~4 Earth radii (Re), permeated by the geomagnetic field — in other words a space plasma. This was one of the first experiments in which the magnetosphere was explored using ground-based instrumentation. Nowadays, there are numerous networks of a variety of instruments widely distributed over most land masses monitoring the state of the magnetosphere. As well as radars, ionospheric sounders, photometers, and riometers, there are networks of relatively inexpensive magnetometers sensing the state of the geomagnetic field, in particular its variability manifested as ultra-low frequency (ULF) waves. As we will see these waves provide an ideal diagnostic tool for remote sensing the composition and dynamics of the magnetosphere.

A classification scheme for ULF waves based on frequency bands and shown in Table 1 was introduced by Jacobs et al.[2] It is based on frequency band observational phenomenological properties of the waves. Pc denotes continuous trains of waves while Pi indicates the more impulsive signatures seen in high resolution magnetograms. Although various wave types are seen across the boundaries, the general nomenclature is still useful today. In this chapter, we are particularly interested in the Pc3–5 bands, typically driven by sources external to the magnetosphere. The higher frequency Pc1–2 waves are generated by internal wave–particle interactions and will not be considered in detail. The spectral characteristics and amplitudes of the various types of waves, illustrated in Fig. 1 from Campbell,[3] shows the largest amplitude waves are seen at the lowest frequencies. Experimentally, Pc3–5 and Pi1–2 waves are observed on fluxgate magnetometers while the lower amplitude higher frequency Pc1–2 waves are better observed on search coil or induction magnetometers which show a rising amplitude response with frequency.

The Earth's magnetosphere is the region of near-Earth space that is threaded by dipole-like magnetic field lines linked to the Earth and

Table 1.   Classification of ULF waves (after Ref. 2).

Continuous pulsations			Irregular pulsations		
Type	Period (sec)	Frequency	Type	Period (sec)	Frequency
Pc1	0.2–5	0.2–5 Hz	Pi1	1–40	25 mHz–1 Hz
Pc2	5–10	0.1–0.2 Hz	Pi2	40–50	7–25 mHz
Pc3	10–45	22–100 mHz			
Pc4	45–150	7–22 mHz			
Pc5	150–600	1–7 mHz			

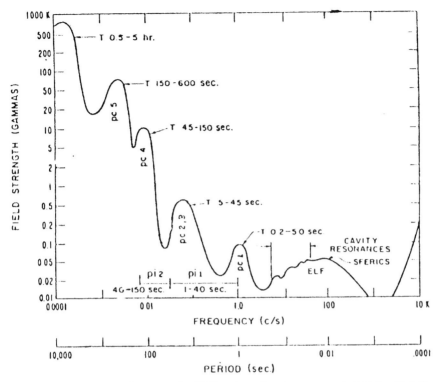

Fig. 1. The spectrum of waves in the ULF regime. Here 1 gamma = 1 nT (after Ref. 3).

in which an electrically charged gas consisting mainly of electrons and protons predominates over the neutral atmosphere. This plasma region, the magnetosphere, has an outer boundary the magnetopause at an altitude of some 60,000 km (∼10 Re), outside of which the turbulent magnetosheath and bow shock interface with the solar wind plasma. The inner boundary of the magnetosphere, below the plasmapause density discontinuity at ∼4 Re and the plasmasphere, is the more dense ionosphere 500 km above the Earth's surface. The magnetospheric cavity, associated regions and current systems embedded in the solar wind plasma flow are illustrated in Fig. 2(a). Figure 2(b) shows a very simple schematic wave model of the magnetosphere. Here Pc3 waves may enter the magnetosphere from the upstream solar wind under favorable conditions. Cavity mode (CCM) or field line resonances of Pc3–5 waves may be driven by impulses on the dayside magnetopause, while the Kelvin–Helmholtz instability may drive Pc5 surface waves on the dawn and dusk flanks of the magnetosphere.

(a)

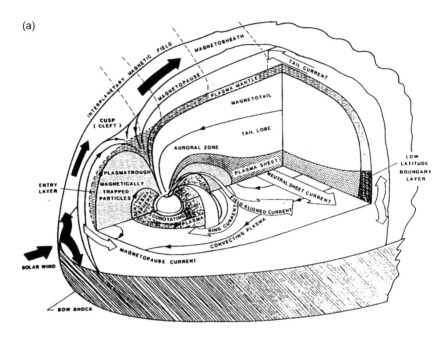

(b)

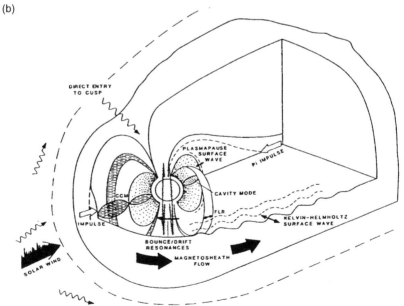

Fig. 2. (a) Cross-section of the Earth's magnetosphere, showing the important plasma regions and current systems (after Ref. 45). (b) A simple schematic showing regions of the magnetosphere populated by various ULF wave types (after Ref. 20).

In addition, the solar wind upstream Pc3 waves may directly enter near the equatorial noon subsolar point or the high latitude cusp regions. Internal magnetosphere wave interactions with charged particles give rise to bounce and drift resonances in the Pc3–5 bands.

This chapter will firstly consider the theory of hydromagnetic wave propagation and resonance in the magnetosphere and define the wave modes, followed by consideration of energy sources for waves, mode coupling, and magnetosphere–ionosphere coupling with respect to ground observations. Finally, it will be shown how the understanding of ULF waves may be used to very inexpensively remote sense the dynamics of the magnetosphere and measure ion mass density. There are many other interesting topics of ULF wave research which are not reported here. These include, for example, high latitude cusp studies, HF radar observations, and radiation belt particle loss and acceleration mechanisms.

2. Wave Equations

In a cold plasma permeated by a magnetic field, as in the terrestrial magnetosphere, ULF waves, commonly referred to as hydromagnetic waves or Alfvén waves, may propagate. They were not well understood until 1942 when Hannes Alfvén predicted their existence and deduced their velocity by combining the classical equations of electrodynamics and hydrodynamics.

Basic wave equations have been derived by many authors e.g., Walker.[4] For simplicity, assume the plasma density ρ is homogenous and that the magnetic field $\mathbf{B}$ is given by a perturbation field $\mathbf{b}$ superimposed on a homogeneous, constant background field $\mathbf{B}_0$, so that $\mathbf{B} = \mathbf{B}_0 + \mathbf{b}$ with $\mathbf{b} \ll \mathbf{B}_0$. The following equations are linearized (by neglecting quadratic terms) and perfect electrical conductivity is assumed.

Then Ohm's law requires that

$$\mathbf{E} = -\mathbf{u} \times \mathbf{B}, \tag{1}$$

where $\mathbf{E}$ is the electric field and $\mathbf{u}$ is the velocity of the plasma and Faraday's law becomes

$$\nabla \times (\mathbf{u} \times \mathbf{B}) = \frac{\partial \mathbf{B}}{\partial t}. \tag{2}$$

The current density $\mathbf{j}$ due to the magnetic disturbance, according to Maxwell's first equation, is given by

$$\mathbf{j} = \frac{1}{\mu_0} \nabla \times \mathbf{b}. \tag{3}$$

Away from the disturbance, where p is the pressure

$$\nabla p = 0. \tag{4}$$

Introducing a Cartesian coordinate system with the z-axis parallel to $\mathbf{B}_0$, it follows that

$$\frac{\partial^2 \mathbf{b}}{\partial t^2} = \frac{\mathbf{B}_0^2}{\rho \mu_0} \frac{\partial^2 \mathbf{b}}{\partial z^2}, \tag{5}$$

$$\frac{\partial^2 \mathbf{u}}{\partial t^2} = \frac{\mathbf{B}_0^2}{\rho \mu_0} \frac{\partial^2 \mathbf{u}}{\partial z^2}. \tag{6}$$

The wave equations (5) and (6) show that the magnetic disturbance $\mathbf{b}$, and the associated velocity of the plasma $\mathbf{u}$, propagate with the Alfvén speed

$$V_A = \frac{\mathbf{B}_0}{\sqrt{\mu_0 \rho}}. \tag{7}$$

Since the Alfvén speed is only a function of the ambient magnetic field strength and the plasma density, it is straightforward to estimate its variation within the magnetosphere, given an appropriate plasma distribution and a realistic geomagnetic field model. This will be considered in Sec. 4.

3. Modes of Propagation

Using a more comprehensive treatment[5,6] it can be shown that three distinct wave modes exist:

(i) The Alfvén mode, which has $\mathbf{u}$ and $\mathbf{b}$ parallel to $\mathbf{k} \times \mathbf{B}_0$, where $\mathbf{k}$ is the wave vector shown in Fig. 3(b). Thus, the magnetic perturbation and the associated movement of the plasma are perpendicular to $\mathbf{B}_0$, while $\mathbf{k}$ has a component parallel to $\mathbf{B}_0$. Since $\mathbf{E}$, $\mathbf{B}_0$, and $\mathbf{k}$ are coplanar, the Poynting vector $\mathbf{S} = \mathbf{E} \times \mathbf{b}$ is parallel to $\mathbf{B}_0$. Therefore, the Alfvén mode is also referred to as the transverse or guided mode. Since $\mathbf{j} \cdot \mathbf{B} \neq 0$, the Alfvén mode carries a finite field aligned current. From the dispersion relation

$$\omega = \pm k V_A \cos \theta, \tag{8}$$

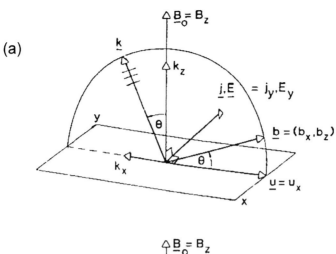

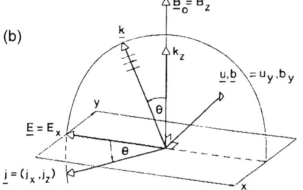

Fig. 3.   Cold plasma hydromagnetic wave modes. (a) Fast mode. (b) Alfvén mode (after Ref. 20).

where θ is the angle between $\mathbf{k}$ and $\mathbf{B}$, it can be seen that the phase velocity $V_A \cos \theta$ is equivalent to motion of the phase fronts in the direction of $\mathbf{B}_0$ with speed V_A.

(ii) The fast mode, which has $\mathbf{j}$ and $\mathbf{E}$ parallel to $\mathbf{k} \times \mathbf{B}$, while $\mathbf{u}$, $\mathbf{b}$, $\mathbf{k}$, and $\mathbf{B}_0$ are coplanar (Fig. 3(a)). Thus, the fast mode does not carry a field aligned current but can transmit pressure variations. Therefore, the fast mode is also referred to as the compressional or magnetosonic mode. The Poynting vector $\mathbf{E} \times \mathbf{b}$ is parallel to $\mathbf{k}$, with energy flow in the direction of propagation. The dispersion relation

$$\omega = \pm k V_A \qquad (9)$$

shows that the phase velocity is independent of the direction of propagation. In the case where the velocity of the oscillating plasma particles and the direction of propagation are parallel to the ambient field $\mathbf{B}_0$, there are no magnetic forces acting, and the wave is purely acoustic, propagating with the velocity of sound C_s. If the velocity of the plasma and the direction of propagation are perpendicular to the ambient magnetic field, the total pressure p_t is the sum of the fluid pressure (as in the acoustic case) and the magnetic pressure. The magnetic field then increases the speed of the wave to

$$C_t = \sqrt{\frac{dp_t}{d\rho}} = \sqrt{C_s^2 + V_A^2}. \qquad (10)$$

In general, there will be some angle between the direction of propagation and the $\mathbf{B}_0$ field, and the speed of propagation of the fast mode will adopt a value between C_s and C_t, i.e., the fast mode will propagate with an elliptical wavefront.

(iii) The slow mode, which appears in hot plasmas only, i.e., when the plasma pressure is significant compared to the magnetic pressure (e.g., in the ring current or the plasma sheet). The slow mode propagates as a compressional wave which is guided along the ambient field with a phase speed comparable to the particle thermal speed, resulting in effective Landau damping. In contrast to the fast mode, where plasma and magnetic pressure oscillate in phase and their effects reinforce each other, in the slow mode they oscillate in anti-phase so their effects tend to cancel.

Typical parameters for the propagation of hydromagnetic waves in the middle magnetosphere in the Pc4–5 band are $B = 100\,\mathrm{nT}$ and $\rho = 5 \times 10^6\,\mathrm{m}^{-3}$ giving $V_A = 10^6\,\mathrm{ms}^{-1}$, $f = 10\,\mathrm{mHz}$ and $\lambda = 10^5\,\mathrm{km} = {\sim}16\,\mathrm{Re}$. Thus, the wavelength is of the scale size of the magnetosphere which suggests waves are more likely to be observed as resonances rather than propagating waves.

4. Hydromagnetic Waves in a General Axisymmetric Dipole Field Geometry

A pure Alfvén wave can only be generated in a uniform magnetic field and plasma. Consequently, the Alfvén and fast modes are generally coupled in the magnetosphere because of inhomogeneities in the magnetic field and the plasma distribution. The coupled wave equations describing the propagation of hydromagnetic disturbances in a general axisymmetric field, e.g., dipole field, have been derived by Dungey.[7,8]

$$\left(\omega^2 \mu_0 \rho - \frac{1}{r}(\mathbf{B} \cdot \nabla) r^2 (\mathbf{B} \cdot \nabla) \right) \left(\frac{u_\phi}{r} \right) = \omega m \left(\frac{\mathbf{B} \cdot \mathbf{b}}{r} \right), \tag{11}$$

$$\left(\omega^2 \mu_0 \rho - r B^2 (\mathbf{B} \cdot \nabla) \frac{1}{r^2 B^2} \mathbf{B} \cdot \nabla \right) (r E_\phi) = i \omega B^2 (\mathbf{B} \times \nabla)_\phi \left(\frac{\mathbf{B} \cdot \mathbf{b}}{B^2} \right), \tag{12}$$

$$i \omega \mathbf{B} \cdot \mathbf{b} = \frac{1}{r}(\mathbf{B} \times \nabla)_\phi (r E_\phi) - i m B^2 \frac{u_\phi}{r}, \tag{13}$$

where a variation of the form $(im\o - i\omega t)$ was assumed. These equations have not been solved analytically except for the special cases outlined below.

(i) *The axisymmetric toroidal mode* $(m = 0)$: This describes a transverse oscillation where the electric field is purely radial and the magnetic field and velocity perturbations are azimuthal. Complete magnetic L shells decouple and perform torsional oscillations with the ΔD-component of the geomagnetic field showing a latitude dependent wave frequency. This mode is illustrated in Fig. 4 assuming perfect ionospheric reflection. It is important to note that the toroidal mode shows out of phase ΔD oscillations between hemispheres for the fundamental (and odd) harmonics, whereas the second (and even) harmonics are in phase.

(ii) *The axisymmetric poloidal mode*: This mode describes a compressional (fast mode) wave, corresponding to alternate compression and expansion of the whole magnetosphere. The electric field is azimuthally directed while the magnetic and velocity perturbations remain in the meridian plane. The fundamental period is about four times the travel time for a fast mode hydromagnetic wave from the ionosphere to the equatorial magnetopause or plasmapause.[9] Since the fast mode is not guided the whole cavity will resonate with the same period at any location and there is no latitude dependence.

(iii) *The guided poloidal mode* $(m \to \infty)$: If a fast mode wave has a large azimuthal wave number and therefore highly localized in longitude, then each meridian plane is decoupled from its neighbor and the wave can be guided along geomagnetic field lines. This frequency will exhibit a distinct latitudinal dependence.[10] This mode is illustrated on the

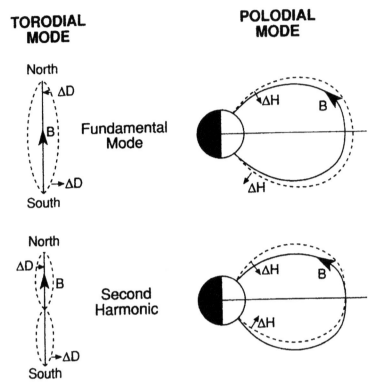

Fig. 4. Schematic showing the two lowest frequency toroidal and poloidal mode oscillations (after Ref. 43).

right-hand side of Fig. 4. Here the ΔH oscillations fundamental (and odd) harmonics are in phase and the second (and even) harmonics are out of phase.

5. ULF Wave Propagation and Resonance

5.1. *Standing Alfvén wave propagation in a realistic magnetosphere*

The formulation of a wave equation applicable to a general geomagnetic field model is given by Singer *et al.*,[11] using the Olsen–Pfitzer field model. Here, two adjacent field lines are separated by a distance δ_α in a plane normal to the field direction. At any other position on the field line, define h_α by requiring the separation to be $h_\alpha \delta_\alpha$. If the normal vector to the field lines is $\boldsymbol{\alpha}$ then for small plasma displacement in the $\boldsymbol{\alpha}$ direction ξ_α, the wave equation is

$$\frac{\partial^2}{\partial s^2}\left(\frac{\xi_\alpha}{h_\alpha}\right) + \frac{\mathrm{d}}{\partial s}\left(\frac{\xi_\alpha}{h_\alpha}\right)\frac{\mathrm{d}}{\partial s}\ln(h_\alpha^2 \mathbf{B}_0) + \frac{u_0 p\omega^2}{\mathbf{B}_0^2}\left(\frac{\xi_\alpha}{h_\alpha}\right) = 0, \qquad (14)$$

where $\mathbf{B}_0$ is the time independent geomagnetic field, s is the distance along the field line, p is the plasma density and ω is the frequency. Singer *et al.*[11] noted a number of approximations were made in this derivation. It is a one-dimensional equation and the azimuthal toroidal mode and radial poloidal modes are solved independently and consideration of mode coupling is not possible. Also field aligned current effects are neglected and the ionospheric boundary conditions must be defined. In spite of these limitations, Eq. (14) may be solved for various plasma density distribution models and realistic geomagnetic field models. For example Waters *et al.*[12] used the T01 magnetic field model[13] and a power law plasma density distribution to study field line resonances and determine plasma mass densities.

If the ULF waves we observe on the ground or satellites are standing waves then they should show the classical amplitude and phase response of a resonance. This is very well illustrated in Fig. 5 where the resonance structure observed in the ionospheric electric field of a 4 mHz toroidal mode wave by the STARE radar shows an amplitude peak of 1° latitude width. The corresponding phase changes by $\sim$180° over this width, and is nearly constant at higher and lower latitudes.

Observations of toroidal and poloidal mode waves have been mainly seen in data from the AMPTE and CRRES satellites. These were high inclination elliptically orbiting satellites providing radial passes through the magnetosphere at all local times. Results from a CRRES study of Pc5 (1–7 mHz) waves by Hudson *et al.*[14] are shown in Fig. 6. Here, the toroidal mode dominates on the dawn and dusk flanks of the magnetosphere, inside geostationary orbit. The poloidal mode, including compressional oscillations, occurs mainly along the dusk side, consistent with high azimuthal wave number excitation by the injection of ring current ions near midnight and on the dusk side.

5.2. *Cavity waveguide modes*

In the 1980s it was suggested by Kivelson and Southwood[15] and Allan *et al.*[16] that ULF waves could be driven by radially standing

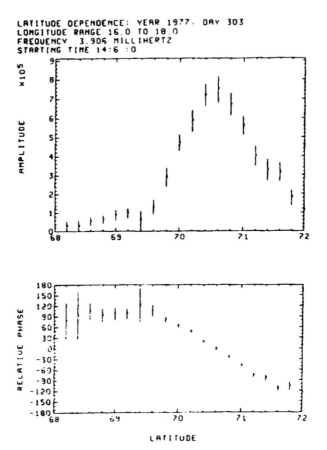

Fig. 5. STARE radar observations of a Pc5 ULF wave field line resonance amplitude and phase response (after Ref. 31).

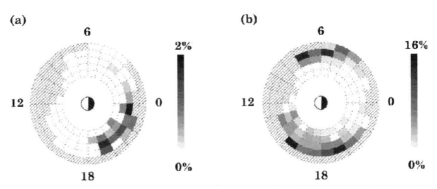

Fig. 6. Occurrence rates of Pc5 ULF waves observed by CRRES. (a) Poloidal mode waves. (b) Toroidal mode waves (after Ref. 14).

compressional disturbances, resulting in magnetospheric cavity mode theory. The frequency spectrum of the cavity modes is structured by the magnetospheric cavity ringing at its natural frequencies. The cavity model was extended by Harrold and Samson[17] and others with the introduction of a waveguide model, with the cavity extending down the tail. Here, the modes propagate anti-sunward down the waveguide with their natural characteristic frequencies. These waveguide modes may be driven by either impulses in the solar wind near the subsolar point or by instabilities on the flanks. The model is illustrated schematically in Fig. 7, from Mann and Wright,[18] where the upper panel, relating to moderate solar wind speeds leads to stagnation over the dayside, and a leaky boundary. With higher speeds across the flanks the boundary becomes perfectly reflecting. For a fast solar wind, shown in the lower panel, the leaky and perfectly reflecting boundaries are protracted toward local noon. Here, the flanks have a fast flow speed which results in an over-reflecting boundary. This means a fast mode inside the magnetosphere and incident on the magnetopause will be reflected with increasing amplitude and therefore energy. The additional energy may be provided by Kelvin–Helmholtz shear flow instability at the boundary.

6. Drift and Bounce Resonance

Southwood *et al.*[19] and Allan and Poulter[20] showed that energetic charged particles that satisfy the condition

$$\omega - m\omega_d = N\omega_b, \tag{15}$$

where N is positive or negative integer, including zero, will resonate with the wave. This represents a corresponding resonance between the phase velocity components of the wave in the field aligned and azimuthal directions, and the bounce and drift velocities of the particles with energies that give the required frequencies. Such particles see a constant wave electric field and can therefore be accelerated or decelerated. The resonance contribution in Eq. (5) is only satisfied by a small fraction of the overall velocity distribution. However, the wave–particle interaction transfers sufficient energy to grow or damp wave modes, depending on the sign of the radial gradient of the velocity distribution function.

This is shown in Fig. 8, where the helical path shows a charged particle bouncing on a field line with **B** increasing at each end (mirroring), while the particle also drifts in azimuth y due to the **B** gradient and curvature

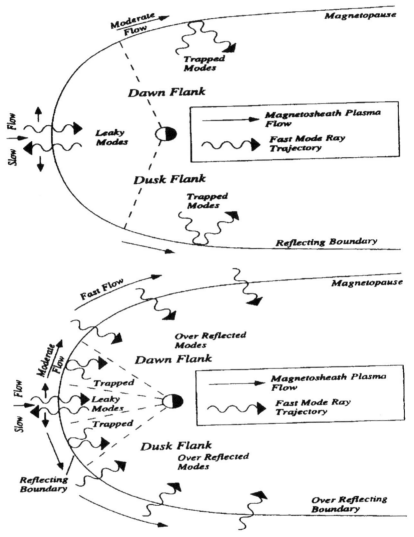

Fig. 7. Waveguide modes driven by solar wind impulses or instability at the flanks.
(a) Moderate solar wind speeds. (b) Fast solar wind (after Ref. 18).

effects. A bounce resonance results when a multiple of the bounce frequency, ω_b matches the eigenfrequency of an even mode field line resonance. If resonance has an appropriate azimuthal wavelength λ_y the drift frequency ω_d is also in resonance by Eq. (15).

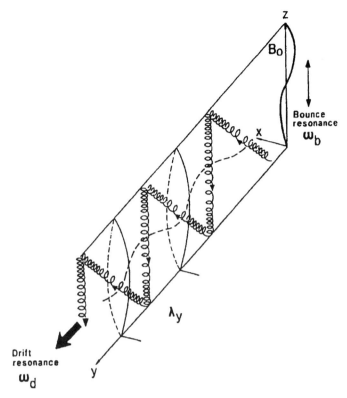

Fig. 8. Schematic illustration of bounce and drift resonances. The helical path with changing pitch describes a charged particle bouncing along a field line with increasing magnetic field at each end, giving a magnetic mirror effect (after Ref. 20).

7. Mode Coupling in the Magnetosphere

A mechanism capable of transferring solar wind wave energy from the magnetopause to deep into the magnetosphere and middle and low latitude ground locations involves coupling of the fast mode wave to the field line resonance. Figure 9(a) shows a meridional cut schematic diagram of the important locations for a fast mode wave propagation in the equatorial plane from the magnetopause at x_b. At the turning point x_t the incoming wave is reflected but some energy propagates across an evanescent region to location x_r where its frequency matches that of the natural field line resonance at x_r. This generates an Alfvén mode where the field aligned phase velocity of the incoming fast mode matches that of the Alfvén wave. Figure 9(b) shows an incoming fast mode wave with non-zero k_y and k_z in

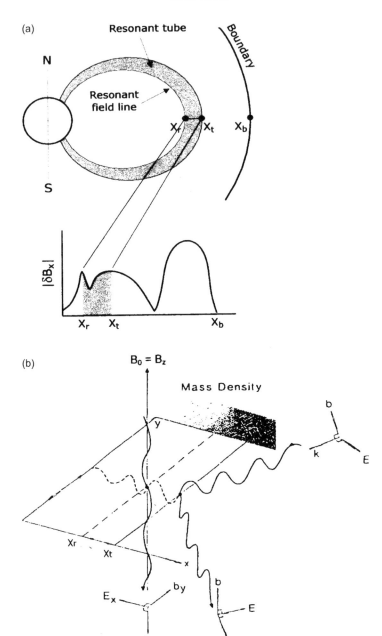

Fig. 9. (a) A meridional cut showing the turning point and resonant point for an incoming fast mode (after Ref. 46). (b) A three-dimensional schematic showing the mode coupling of the fast mode to a field line resonance (after Ref. 20).

a box model with a uniform magnetic field.[20] Note that the plasma mass density decreases with decreasing x is not realistic but has been chosen because the magnetic field is constant with x to provide a realistic increase in Alfvén velocity with decreasing x. If the fast mode wave is sustained, energy will accumulate in a narrow region around x_r.

8. Energy Sources

8.1. *Transfer of ULF wave power into the magnetosphere*

Dungey[7] was the first to suggest that ULF wave energy may be transmitted into the magnetosphere by surface waves excited on the magnetopause by the magnetosheath flow. This is the Kelvin–Helmholtz instability, analogous to a wind over water scenario. Figure 10 illustrates the phenomenon which occurs at the dawn and dusk flanks. It is also seen that waves travel westward pre-noon and eastward post-noon while Pc5 wave polarization is left hand in the morning and right hand in the afternoon. These results have been confirmed by a number of ground studies.[21] The same properties are seen at middle and low latitudes, although the mechanism may be different.

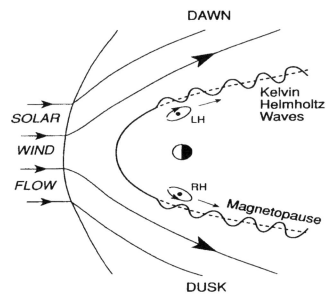

Fig. 10. Equatorial plane schematic illustrating the solar wind flow past the magnetopause. High velocity at the flanks initiates the Kelvin–Helmholtz instability and fast mode wave energy is transmitted into the magnetosphere (after Ref. 43).

Recently, there has been a number of experimental studies indicating the generation of Pc5 ULF waves in the magnetosphere by external solar wind energy input. Villante *et al.*[22] studied the coherence between perturbations seen in the solar wind density and velocity and the interplanetary magnetic field with Pc5 waves seen at geosynchronous orbit and on the ground. Some of these results are shown in Fig. 11 where the coherency between the various parameters is shown in the left panels

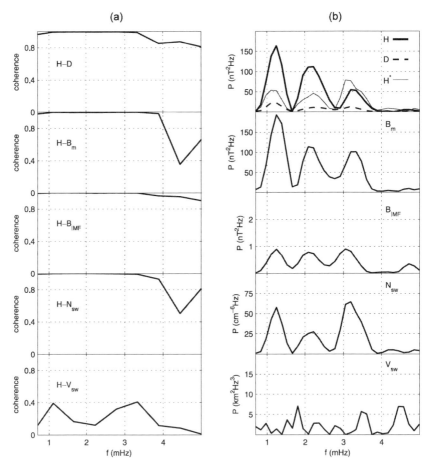

Fig. 11.   (a) Coherence between H component on the ground with different parameters as labelled. (b) The corresponding power spectra. The parameter H^* predicts the effects of the variable solar wind pressure and magnetospheric currents and is similar to H, and Bm is the magnetospheric field magnitude at geostationary orbit.[47] (after Ref. 22).

and the spectra of the corresponding perturbations in the right panels. It is remarkable that common frequencies and harmonics are seen in bands centered on 1.3, 2.3, and 3.2 mHz. This asks the question of whether cavity or waveguide modes can be directly driven by the solar wind or are they excited by cavity waveguide modes. At present both seem to be effective and the answer is not clear.

Another example where solar wind pressure and velocity enhancements appear to excite discrete Pc5 waves has been presented by Rae *et al.*[23] An interval of long lasting narrowband Pc5 waves seen in the dusk sector during a geomagnetic storm recovery phase was observed by the Cluster, Polar, GOES and LANL satellites in the magnetosphere, HF radars in the ionosphere and the CANOPUS magnetometer array on the ground. There were no monochromatic dynamic pressure or solar wind velocity variations observed at this time and it was concluded that the waves were a result of an excited waveguide mode, possibly by the Kelvin–Helmholtz instability or via over-reflection at the dusk magnetopause. The results are summarized in Fig. 12. Panel (a) shows an absence of the 1.5 mHz monochromatic oscillation in upstream solar wind data. The oscillations are seen in panels (b) to (g), at geosynchronous orbit, both in the magnetic field and particles, and also by SuperDARN HF radars and ground magnetometers.

8.2. *Energy sources for Pc3 ULF waves*

The higher frequency Pc3 (10–100 mHz) waves have an external energy source attributed to upstream waves in the solar wind.[24] These are ion cyclotron waves located upstream of a parallel shock region, which propagate into the magnetosheath and through the magnetosphere and plasmasphere to the ground by compressional cavity mode fast wave processes previously described in Secs. 5.2 and 7. Wave amplitudes in the magnetosphere maximize when the interplanetary magnetic field is nearly parallel or anti-parallel to the Sun–Earth line,[25] and the frequency is given by the empirical relation $f(\text{mHz}) = 6B_{\text{IMF}}(\text{nT})$.[26] Thus, the wave parameters seen on the ground are determined by solar wind and interplanetary magnetic field conditions rather than magnetospheric conditions. Figure 13, showing the meridian and ecliptic planes, illustrates how the upstream Pc3 wave energy enters and is distributed throughout the magnetosphere. It is also important to note

25 Nov 2001

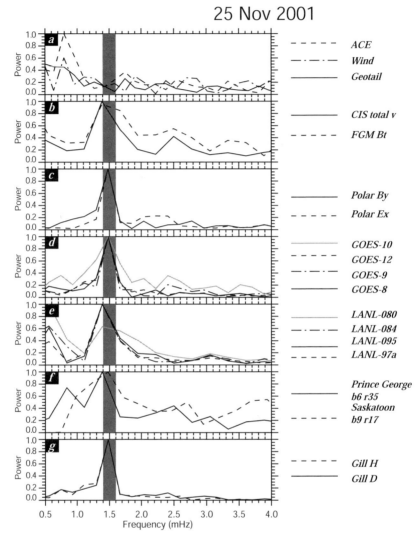

Fig. 12.   Schematic model of the process by which Pc3 waves may be propagated into the magnetosphere from the upstream region (after Ref. 23).

that the magnetosphere–plasmasphere system may act as a spatial filter, and wave diffraction may occur around the Earth. Radar and magnetometer observations supporting this scenario have been presented by Ponomarenko et al.[27] and Howard and Menk.[28]

Ecliptic plane

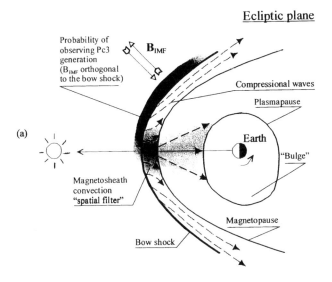

Magnetic meridian plane

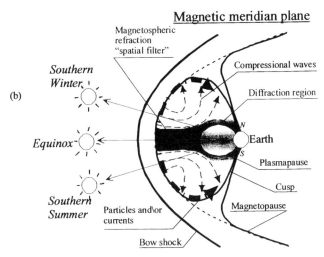

Fig. 13. Pc3 ULF wave entry and propagation in the magnetosphere. (a) Ecliptic plane. (b) Meridional plane (P. V. Ponomarenko, unpublished).

9. ULF Wave Propagation Through the Ionosphere

For ULF wave energy to be observed on the ground it must pass through the ionosphere, and this region modifies some wave characteristics. The ionosphere has a finite anisotropic conductivity perpendicular to the magnetic field so the assumption of infinite conductivity and the reflection of Alfven waves is an approximation. However, the fast mode wave is not significantly affected by the ionosphere and the boundary condition comes directly from Maxwell's induction equation. Application of this boundary condition has been considered by Yeoman et al.[29] The situation for an incident oblique Alfven wave is not simple since this wave carries a field aligned current and must also show zero vertical current at the base of the ionosphere. The situation for an idealized field aligned current system is explained by Allan and Poulter[20] and is illustrated in Fig. 14 for one point on the wave cycle. The current $J_\parallel$ is closed by Pedersen current J_p in the E region with a corresponding electric field E_x in the ionosphere. A Hall current J_H is generated by E_x. The $(J_\parallel, J_p)$ system is approximately solenoidal, so very little magnetic field leaks out. J_H has a magnetic field seen as a b_x component at the ground. The resulting effect is that the

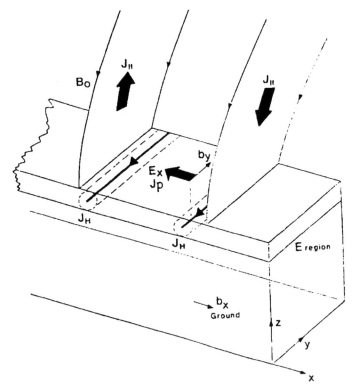

Fig. 14.   Diagram illustrating the field aligned current system of an Alfvén wave incident on a conducting ionosphere (after Ref. 20).

original component of the Alfven wave above the ionosphere b_y appears as b_x on the ground. Thus, the signal has been rotated by $90°$ through the ionosphere. Here the azimuthal b_y toroidal field line resonance component above the ionosphere will be observed as a north–south b_x component on the ground. Ionospheric Joule dissipation dictates a reflection coefficient less than unity. The $90°$ rotation of the polarization ellipse major axis was predicted by Hughes[30] and confirmed by Walker *et al.*[31] using radar data.

10. Heavy Ion Effects on Field Line Resonance

An Alfven wave propagating through the magnetosphere has a velocity which is determined by the magnetic field strength and the plasma density along the field line path. For a field line resonance the resonant frequency will be dependent on the plasma composition, and in particular its constituents. The heavy ions He^+ and O^+ play an important role and Fraser *et al.*[32] showed that the mass density composition can have profound effects on the frequency of field line resonances. It will be seen in the next section that this has important ramifications for the application of ULF wave plasma diagnostic techniques. Figure 15 shows how field line resonant frequencies are modified by the presence of the oxygen torus which is often seen during the geomagnetic storm recovery phase. Panel (a) shows the mass density increases dramatically over $L = 2.5–4.5$. It has the effect of moving the H^+ or (H^+, He^+) plasmapause from $L = 2.5$ to $L = 4.5$. Panels (b) and (c) show the effect this has on the toroidal wave resonant frequency. It is therefore important to use a heavy ion mass loaded density when studying Alfven wave propagation in the plasmasphere–magnetosphere. The use of electron density or an equivalent single ion proton plasma population implying a mass loading of unity is not generally correct. A recent comprehensive review of mass loading has been given by Denton.[33]

11. Remote Sensing the Magnetosphere Using Ground-Based Observations of ULF Waves

A single dominant and discrete driven field line resonance may be identified over a latitudinal chain of stations. This was based on Dungey's original ideas[7] and expanded by Tamao.[34] The detailed properties relating to field line resonance and associated polarization properties, which show a polarization reversal across the resonance were predicted by Southwood[35]

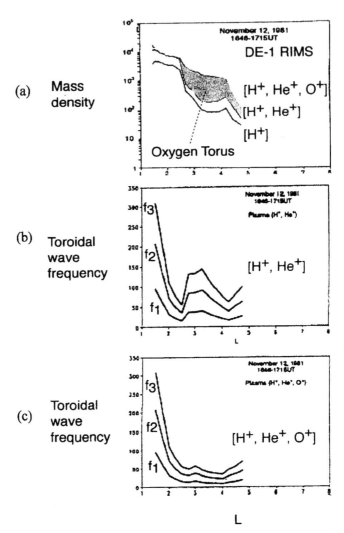

Fig. 15.   Heavy ion mass loading effects on Alfvén wave field line resonances. (a) Satellite ion density data show mass density for various combination of heavy ions, including the oxygen torus. (b) Toroidal mode eigenfrequencies for three harmonics in a (H^+, He^+) plasma. (c) The effects of adding the oxygen to mass density on eigenfrequencies (after Ref. 32).

and Chen and Hasegawa.[36] If a number of driven resonances are present or decoupled L-shells are oscillating at their natural eigenfrequencies due to stimulation from a broadband source, then unambiguous identification of toroidal mode resonance is impossible. Also the resonance frequency observed may not be located at a specific station latitude due to spatial integration effects[37] or wave mode coupling. These difficulties led Baransky *et al.*[38] to propose the amplitude gradient method for measuring the resonant frequency of a field line located midway between two north–south ground stations spaced at typically <100 km at $L \sim 3$. In Fig. 16 the top panel shows the resonance amplitude response of the H component from each station. Since they are at slightly different latitudes they will show different resonant frequencies ω_S and ω_N. The zero-crossing point of the amplitude ratio shown in the middle panel identifies the resonant frequency ω_R of the field line at the latitude of the midpoint between the two stations. Being an amplitude measurement technique this is prone to noise which may affect the location of the zero crossing point, and often more than one zero crossing is identified. Subsequently, Waters *et al.*[39] showed that the cross-phase spectral response between the two station phases showed a peak at the resonant frequency ω_R and this more reliable method is shown in the bottom panel. Reliability of the technique is seen in Fig. 17, where 20 consecutive days of field line resonance frequency data are plotted for a station located at $L = 2.7$. Resonances are seen on 18 of the 20 days, many showing up to four harmonics.

It soon became clear that this method could be developed into a remote sensing tool to probe the plasmasphere–magnetosphere density. The eigenperiod T of a particular field line is given by

$$T = 2 \int_{-\ell/2}^{+\ell/2} \frac{1}{V_A} ds, \qquad (16)$$

where ℓ is the length of the field line, V_A the Alfvén velocity, and ds an element along the field line. Since V_A includes the geomagnetic field $\mathbf{B}$ and mass loaded plasma density ρ, radially dependent models must be provided for these quantities. $\mathbf{B}$ fields used include dipole and Tsyganenko models, while plasma density models may be a simple power law radial distribution with exponent ranging from $n = 0$ to -6, with typically -3 or -4 used. More complicated models have been suggested by Denton.[33] A simple inversion of Eq. (16) including these models, allows the equatorial mass density to be determined.

Cross-phase identification of individual field line eigenfrequencies coupled with the inversion of Eq. (16) has proven to be a powerful and inexpensive method of monitoring the mass loaded ion density in the plasmasphere and the plasmapause. Some results from Menk *et al.*[40] are shown in Fig. 18. The top panel shows the distribution of toroidal mode

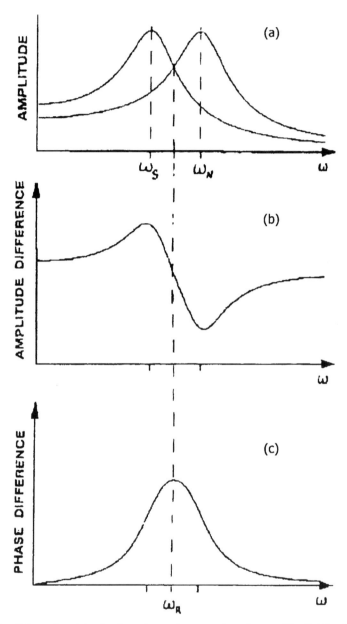

Fig. 16. Schematic plots for two damped resonant systems with slightly different eigenfrequencies. (a) The amplitude response $A(\omega)$ in each case. (b) The meridional amplitude difference $A(\omega_S) - A(\omega_N)$. (c) The meridional cross-phase difference $\varphi(\omega_S) - \varphi(\omega_N)$. In (b) the resonant frequency ω_R is determined by the condition $A(\omega_S) - A(\omega_N) = 0$, and in (c) where $\varphi(\omega_S) - \varphi(\omega_N)$ is a maximum (after Ref. 39).

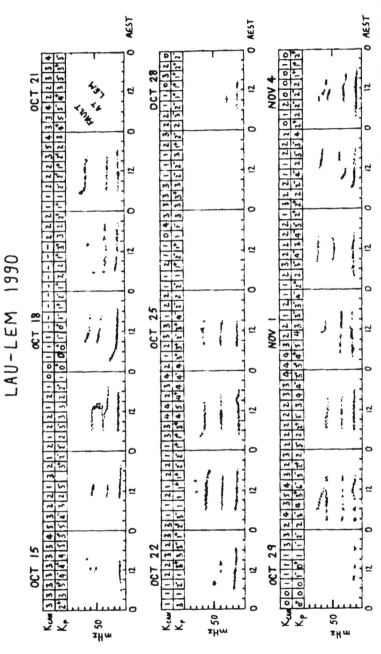

Fig. 17. A series of 21 consecutive days showing cross-phase dynamic spectra between Launceston and Lemont at $L = 2.7$ (after Ref. 44).

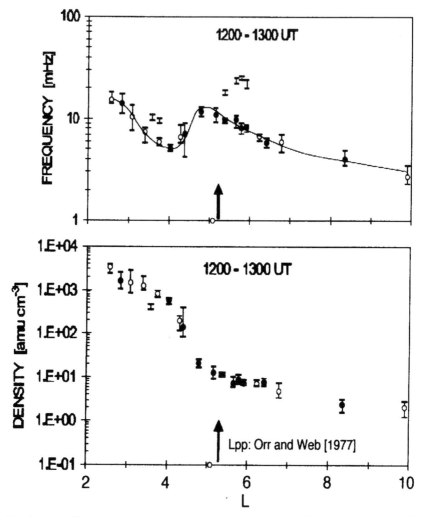

Fig. 18. (a) Variation of resonant frequency with latitude. (b) Mass density profile (after Ref. 40).

field line resonance frequencies with L value determined from the IMAGE magnetometer array in Scandinavia. The lower panel shows the mass loaded plasma density with a distinct plasmapause located at $L \sim 4.5$. This ground based monitoring technique may also be used to study plasmasphere dynamics. In particular Grew et al.[41] has shown that the evolution of plasmapause notches and attached radial drainage plumes, identified in

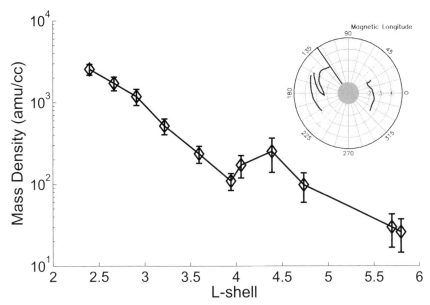

Fig. 19. Mass density profile showing evidence for a plasma drainage plume around $L = 4.3$. The ground magnetometer chain longitude is shown by the radial dashed line in the polar plot on the top right (after Ref. 41).

IMAGE-EUV satellite data by Sandel[42] and others may be tracked in longitude. With respect to notches, a sub-corotation rate of 81% was identified, in agreement with IMAGE-EUV satellite observations. Figure 19 shows the observation of a plasma drainage plume in the radial density profile over $L = 4.0$–4.7, determined from IMAGE magnetometer data. The passage of the drainage plume, seen at ~150° magnetic longitude in the polar plot in the right top corner of Fig. 19, through the plasma trough relates to the plume density enhancement seen in the magnetometer data. It is however important to note that ULF waves must necessarily be present in order to make these measurements and they are typically seen only on the dayside between dawn and dusk, as seen in Fig. 17. Remote sensing the dynamics of the plasmasphere–magnetosphere closed field line environment simultaneously at many longitudes has the potential to provide continuous observations, currently unavailable with satellites. This is one of the aims of the Ultra Large Terrestrial International Magnetometer Array (ULTIMA) (http://www.serc.kyushu-u.ac.jp/index_e.html) international collaborative effort that will be realised over the next few years.

12. Summary

Ultra-low frequency (ULF) waves in the 1–100 mHz band are ubiquitous in the plasmasphere and magnetosphere. They are a manifestation of hydromagnetic wave activity generated by physical processes resulting from the interaction of the solar wind with the Earth's magnetosphere. The almost pure sinusoidal signature of ULF waves when seen on the ground by magnetometers and in space by satellites suggests a resonance phenomenon. This is not unexpected since the size of the magnetospheric cavity is of the same order as the wavelength of the ULF waves. The magnetopause, plasmapause, and ionosphere provide convenient boundaries for wave reflection and transmission. Cavities bounded by these surfaces produce toroidal, poloidal, and fast mode resonances from propagating modes. The Alfven toroidal wave mode is field aligned, resulting in field line resonance while the fast mode propagates isotropically and may establish cavity or waveguide resonances in the magnetosphere. These wave modes couple in the magnetosphere. The presence of a conducting ionosphere has a profound affect on the transmission and polarization properties of the waves seen on the ground.

The propagation and resonance characteristics of ULF waves can be used as diagnostic tools to determine important information on the topology of the dynamic plasmasphere and magnetosphere at all latitudes, from both spatial and temporal perspectives, and complement and extend satellite results. In the future it is envisaged that latitudinal chains of magnetometers spaced in longitude will simultaneously monitor the plasmasphere/magnetosphere system, in particular the cold mass loaded plasma.

References

1. L. R. O. Storey, *Phil. Trans. Roy. Soc.* **A246**(908) (1953) 113.
2. J. A. Jacobs, K. Kato, S. Matsushita and V. A. Troitskaya, *J. Geophys. Res.* **69** (1964) 180.
3. W. H. Campbell, *Proc. IRE* **51** (1963) 1337.
4. A. D. M. Walker, *Magnetohydromagnetic Waves in Geospace: The Theory of ULF Waves and Their Interaction with Energetic Particles in the Solar-Terrestrial Environment* (Ser. Plasma Phys., Inst. Phys. London, 2003).
5. J. W. Dungey, *Cosmic Electrodynamics* (Cambridge University Press, 1958).
6. H. Alfven and C. G. Falthammar, *Cosmical Electrodynamics* (Oxford University Press, 1963).
7. J. W. Dungey, *Ionos. Res. Lab. Penn. State Univ., Sci. Rep.* **69** (1954).

8. J. W. Dungey, *Geophysics, The Earth's Environment*, eds. C. De Witt, J. Hieblot and A. Leleau (Gordon and Breach, 1963).
9. J. W. Dungey and D. J. Southwood, *Space Sci. Rev.* **10** (1970) 672.
10. H. R. Radoski, *J. Geophys. Res.* **72** (1967) 418.
11. H. J. Singer, D. J. Southwood, R. J. Walker and M. G. Kivelson, *J. Geophys. Res.* **86** (1981) 4589.
12. C. L. Waters, J. C. Samson and E. F. Donovan, *J. Geophys. Res.* **101** (1996) 24737.
13. N. A. Tsyganenko, *J. Geophys. Res.* **107** (2002) 1176, doi: 10.1029/2001JA000220.
14. M. K. Hudson, R. E. Denton, M. R. Lessard, E. G. Miftakhova and R. R. Anderson, *Ann. Geophys.* **22** (2004) 289.
15. M. G. Kivelson and D. J. Southwood, *Geophys. Res. Lett.* **12** (1985) 49.
16. W. Allan, S. P. White and E. M. Poulter, *Geophys. Res. Lett.* **12** (1985) 287.
17. B. G. Harrold and J. C. Samson, *Geophys. Res. Lett.* **19** (1992) 1811.
18. I. R. Mann and A. N. Wright, *Geophys. Res. Lett.* **26** (1999) 2609.
19. D. J. Southwood, J. W. Dungey and R. J. Etherington, *Planet. Space Sci.* **17** (1969) 349.
20. W. Allan and E. M. Poulter, *Rep. Prog. Phys.* **55** (1992) 533.
21. S. T. Ables, B. J. Fraser, C. L. Waters, D. A. Neudegg and R. J. Morris, *Geophys. Res. Lett.* **25** (1998) 1507.
22. U. Villante, P. Francia, M. Vellante, P. Di Giuseppe, A. Nubile and M. Piersanti, *J. Geophys. Res.* **112** (2007), doi: 10.1029/2006JA011896.
23. I. J. Rae, E. F. Donovan, I. R. Mann, F. R. Fenrich, C. E. J. Watt, D. K. Milling, M. Lester, B. Lavraud, J. A. Wild, H. J. Singer, H. Reme and A. Balogh, *J. Geophys. Res.* **110** (2005), doi: 10.1029/2005JA011007.
24. M. M. Hoppe and C. T. Russell, *J. Geophys. Res.* **88** (1983) 2021.
25. E. W. Greenstadt and J. V. Olson, *J. Geophys. Res.* **81** (1976) 5911.
26. V. A. Troitskaya, T. A. Plyasove-Bakunina and A. V. Gul'elmi, *Dokl. Akad. Nauk, USSR* **197** (1971) 1312.
27. P. V. Ponomarenko, F. W. Menk, C. L. Waters and M. D. Sciffer, *Ann. Geophys.* **23** (2005) 1271.
28. T. A. Howard and F. W. Menk, *J. Geophys. Res.* **110** (2005), doi: 10.1029/2004JA010417.
29. T. K. Yeoman, M. Lester, D. Orr and H. Luhr, *Planet. Space Sci.* **38** (1990) 1315.
30. W. J. Hughes, *Planet. Space Sci.* **22** (1974) 1157.
31. A. D. M. Walker, R. A. Greenwald, W. F. Stuart and C. A. Green, *J. Geophys. Res.* **84** (1979) 3373.
32. B. J. Fraser, J. L. Horwitz, J. A. Slavin, Z. C. Dent and I. R. Mann, *Geophys. Res. Lett.* **32** (2005), doi: 10.1029/2004GL021315.
33. R. E. Denton, *Magnetospheric ULF Waves: Synthesis and New Directions*, eds. K. Takahashi, P. J. Chi, R. E. Denton and R. L. Lysak, *Geophys. Monograph Ser.*, Vol. 169 (American Geophysical Union, 2006).
34. T. Tamao, *Sci. Rep. Tohoku Univ., Ser. 5* **17** (1965) 43.
35. D. J. Southwood, *Planet. Space Sci.* **22** (1974) 483.

36. L. Chen and A. Hasegawa, *J. Geophys. Res.* **79** (1974) 1033.
37. P. V. Ponomarenko, C. L. Waters, M. Sciffer, B. J. Fraser and J. C. Samson, *J. Geophys. Res.* **106** (2001) 10509.
38. L. N. Baransky, J. E. Borovkov, M. B. Gokhberg, S. M. Krylov and V. A. Troitskaya, *Planet. Space Sci.* **33** (1985) 1369.
39. C. L. Waters, F. W. Menk and B. J. Fraser, *Geophys. Res. Lett.* **18** (1991) 2293.
40. F. W. Menk, I. R. Mann, A. J. Smith, C. L. Waters, M. A. Clilverd and D. K. Milling, *J. Geophys. Res.* **109** (2004), doi: 10.1029/2003JA010097.
41. R. S. Grew, F. W. Menk, M. A. Clilverd and B. R. Sandel, *Geophys. Res. Lett.* **34** (2007), doi: 10.1029/2006GL028254.
42. B. R. Sandel, J. Goldstein, D. L. Gallagher and M. Spasojevic, *Space Sci. Rev.* **109** (2003) 25.
43. W. J. Hughes, *Solar Wind Sources of Magnetospheric Ultra-Low Frequency Waves*, eds. M. Engebretson, K. Takahashi and M. Scholer, *Geophys. Monograph Ser.*, Vol. 81 (American Geophysical Union, 1994).
44. C. L. Waters, Low latitude geomagnetic field-line resonance, PhD thesis, University of Newcastle (1993).
45. D. P. Stern, *Office of Space Science*, NASA (1978).
46. J. C. Samson, B. G. Harrold, J. M. Ruohoniemi, R. A. Greenwald and A. D. M. Walker, *Geophys. Res. Lett.* **19** (1992) 441.
47. N. A. Tsyganenko and D. P. Stern, *J. Geophys. Res.* **101** (1996) 27187.

Advances in Geosciences
Vol. 14: Solar Terrestrial (2007)
Eds. Marc Duldig et al.
© World Scientific Publishing Company

NEW SUNRISE OF SOLAR PHYSICS GALVANIZED BY THE *HINODE* MISSION*

TETSUYA WATANABE

National Astronomical Observatory
2-21-1 Mitaka, Tokyo 181-8588, Japan

HINODE SCIENCE TEAM

Japan Aerospace Exploration Agency
3-1-1 Yoshinodai, Sagamihara, Kanagawa Pref., 229-8511, Japan

The Solar-B mission was successfully launched on 22-Sep-2006 (UT), put into a sun-synchronous polar orbit, and called "*Hinode*," the sunrise in Japanese. All three mission telescopes on board, namely, solar optical telescope (SOT), X-ray telescope (XRT), and extreme ultraviolet imaging spectrometer (EIS) opened their doors about a month after the launch and started the initial observations. Thanks to the sun-synchronous orbit, observations will be possible 24 h a day for about 8 months of the year. The 50 cm diameter SOT is able to obtain a continuous, seeing-free series of diffraction-limited images (0.2–0.3″) in the wavelengths of 388–668 nm, and a wide range of scientific advances are anticipated through this increased capability. XRT is a high resolution grazing incidence telescope, a successor to the very successful SXT instrument on board *Yohkoh*. The EIS utilizes an off-axis parabolic primary and a toroidal diffraction grating in a normal incidence optical layout with multi-layer coating, which enables high reflectance two EUV wavelength ranges, 170–210 Å and 250–290 Å. The current status of the *Hinode* mission and the initial scientific results will be summarized.

1. Introduction; Successful Launch of *Hinode*

The Solar-B mission was successfully launched from Uchinoura Space Center (USC)/JAXA at 18:36:00 on 22-Sep-2006 (UT), and called

---

*Hinode is a Japanese mission developed and launched by ISAS/JAXA, collaborating with NAOJ as domestic partner, NASA and STFC (UK) as international partners. Scientific operation of the Hinode mission is conducted by the Hinode science team organized at ISAS/JAXA. This team mainly consists of scientists from institutes in the partner countries. Support for the post-launch operation is provided by JAXA and NAOJ (Japan), STFC (UK), NASA (USA), ESA, and NSC (Norway)

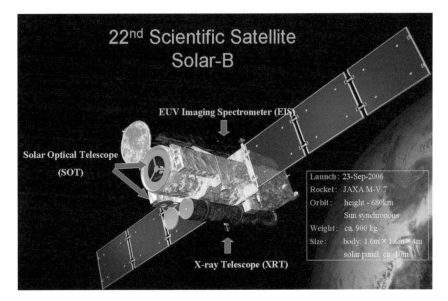

Fig. 1.   The 22nd Japanese Scientific Satellite, "*HINODE*".

"*Hinode*," (Fig. 1) the sunrise in Japanese. The scientific mission objectives are to understand the basic questions of solar physics; (a) the coronal heating mechanism, (b) the origin of strong magnetic fields, and (c) the trigger mechanism of solar flares along with other questions.

It was successfully put into a sun-synchronous polar orbit at an altitude of 680 km above the sea level on 3-Oct-2006. The telescope doors of the three scientific instruments, i.e., Solar optical telescope (SOT), X-ray telescope (XRT), and extreme ultraviolet imaging spectrometer (EIS), were opened about a month after the launch, and their performances in orbit were successfully verified. Then, *Hinode* started initial scientific observations in early December, and a guest-investigation-type program was announced in the "Call for Proposals." These were accepted from 12-Dec-2006 to encourage the world-wide scientific involvement in the *Hinode* observations.[a] All the data taken by the telescopes, as well as those for house-keeping purpose have been immediately open to the public and available[b] since 27-May-2007 with instruction for data users.

---

[a]See: http://solar-b.mtk.nao.ac.jp/operation_e/proposal_e.shtml
[b]See:http://darts.isas.jaxa.jp/hinode/top.do and http://hinode.nao.ac.jp/hsc_e/instruction_e.shtml

2. Instrumentation

The capabilities of three telescopes on board will be very briefly described in this section. Details will be given in Solar Physics.[1–8]

2.1. *Solar optical telescope (SOT)*

The complete SOT instrument consists of the optical telescope assembly (OTA), and the focal plane package (FPP). It is a Gregorian telescope with an effective aperture of 50 cm that is aiming at diffraction-limited performance of OTA with the tip-tilt movable mirror (TCM) system, namely, 0.2–0.3 s at the wavelengths 380–670 nm.

The FPP includes a filtergraph (FG) and a spectrograph (SG). FG takes high spatial resolution images in the 12 wavelength bands, and 3D photospheric magnetograms with narrowband filters (NFI) and broadband filters (BFI). SG concentrates in taking the full set of polarimetric data of the iron (FeI) absorption line at 6302 Å.

2.2. *X-ray telescope (XRT)*

The X-ray telescope (XRT) is a grazing incidence mirror telescope accommodating a $2k \times 2k$ CCD as a detector. The spatial resolution is improved to 1 arcsec, about three times better than that of *Yohkoh*/SXT, and its measurable plasma temperature range is increased to 1–20 million K. The field of view of this instrument is the widest among the three *Hinode* instruments and covers the entire solar disk, when the spacecraft is pointed to the disk center. The electronic systems for automatic exposure control and flare detection are accommodated in the mission data processor (MDP).

2.3. *EUV imaging spectrometer (EIS)*

The major improvement of the EIS telescope throughput was achieved by applying multi-layer coatings to the mirror and the grating used in the EUV wavelength range. The two wavelength bands observed with the EIS instrument are 170–210 Å and 250–290 Å and include the emission lines of HeII and FeVIII–FeXXIV, lines formed in the temperature range $4.7 < \mathrm{Log}\, T < 7.2$.

The plate scale is such that one CCD detector pixel corresponds to 1 arcsec in the spatial direction along the slit and the spectral resolution is

0.00223 Å/pixel. Four kinds of slit/slots, slits of 1 and 2 arcsec(s) and slots of 40 and 266 arcsecs, are accommodated and may be interchanged in a housing near the focal plane of the primary mirror. A total of 25 emission line windows can be selected on two CCDs, dedicated to the shorter and the longer wavelengths, respectively. The fine velocity structure and non-equilibrium nature of solar coronal plasmas will be observed for the first time at EUV wavelengths with this capable instrument.

The fields of view and temperature coverage of the three scientific instruments on board the *Hinode* mission are summarized in Fig. 2.

3. Brief Summary of Initial Results

3.1. *Telescope performance*

During the commissioning phase and in the initial observing period after first light, the on board performance of the telescopes was thoroughly checked and confirmed as had been expected during the calibration phase:

- Images taken by SOT/OTA revealed that the diffraction limited performance of 0.2 arcsec spatial resolution at 5,000 Å is continuously achieved on board.

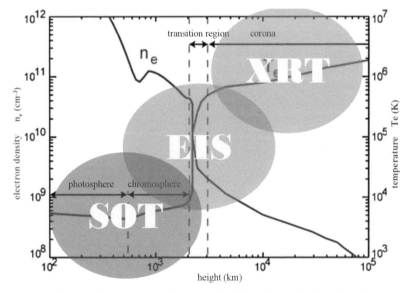

Fig. 2(a).   Temperature coverage of *Hinode* scientific instruments; SOT, XRT, and EIS.

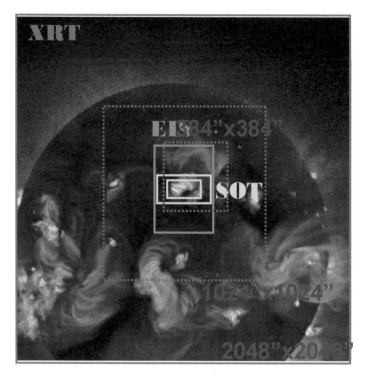

Fig. 2(b).   Fields of view of the telescopes; SOT, XRT, and EIS.

- XRT just starts to reveal loop-like structures of X-ray bright points (XBPs) in the coronal holes, proving its spatial resolution of ∼1 arcsec.
- Comparing the images simultaneously taken by *Hinode*/EIS and SoHO/CDS, an order of magnitude improvement of spatial resolution and sensitivity has been achieved. Numerous unidentified weak emission lines are observed in both of the EIS wavelength ranges.

The improvement of the telescopes' performance can be easily recognized in the panels of Figs. 3(a)–3(c).

The initial 90 days are dedicated to verifying the optical performance of the telescopes, and already in December 2006, observation proposals from all over the world have been widely accepted to encourage collaboration with ground-based facilities and with multiple instruments on board *Hinode*. By September 2007, more than 40 proposals have been accepted in the meetings of science schedule coordinators (SSCs), and carried out on board. *Hinode* experienced the first day-night cycling season from the

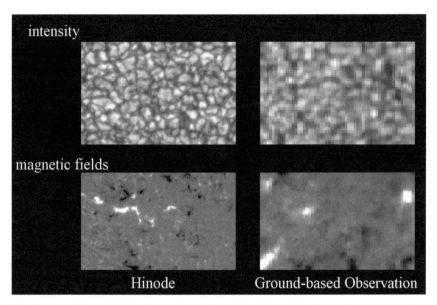

Fig. 3(a). SOT image quality compared with that by ground-based telescopes. The absolute merit of SOT is to provide continuously the best snap shot images taken by the ground-based facilities.

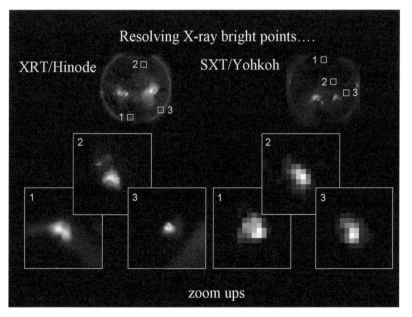

Fig. 3(b). Comparing similar features of the solar corona observed by XRT and SXT/Yohkoh; XRT starts to resolve the structure of XBPs.

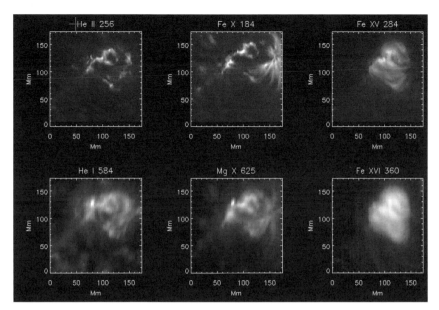

Fig. 3(c). Image quality of EIS is compared with that of CDS/SoHO by choosing emission lines formed at similar temperatures: top panel are images taken by EIS, bottom by CDS.

beginning of May to the middle of August in 2007, and regular night time interruptions of observations occurred with a maximum duration of 20 min per orbit during this period.

3.2. *New discoveries by the* Hinode *mission*

More than 40 papers describing the first results of *Hinode* will be published in the special issue of Publication of Astronomical Society in November 2007.[9] New discoveries and findings during the first 9 month observations of *Hinode* can be itemized as follows:

New Discoveries:

- Strong magnetic fields ($B > 1$ kG) are observed in the polar regions, where the magnetic fields have previously been thought to be weak and vertical.
- Horizontal strong magnetic fields, though transient by nature, are observed in the quiet sun and in the polar region as well. They are numerous.

- Convective collapse, one of the mechanisms to make up strong magnetic fields in the photosphere, could be observed in the diffraction limited magnetograph images.
- Alfvén waves are seen to propagate throughout the photospheres and the chromospheres.
- Jets and/or supersonic flows are also present anywhere in the quiet sun and also in the coronal holes.

First detections and detailed analyses are continuing for the following science topics:

- Collapsing processes in sun spots; cancellations and escapes of magnetic fluxes from sunspots will be studied in detail with continuous magnetograph observations.
- Flows associated with flares will be studied. Inflows/outflows to/from the magnetic reconnection sites, chromospheric evaporations and down flows above flare arcades will be closely investigated with all three instruments.
- Turbulence or turbulent flows at the foot points of coronal loops seem universal in most active regions. This will provide clues to allow identification of the mechanism of coronal heating.
- Temperature structures along with height will be investigated through XRT and EIS by calibrating their diagnostic capabilities.
- Polar jets found in the polar coronal holes will be studied in detail, to understand their contribution to coronal heating, polar plumes, and the fast and slow solar winds.
- Oscillations are seen everywhere in the photosphere. Umbral oscillations in particular will reveal a method for energy transport in strong magnetic fields.
- Numerous interesting phenomena, such as flows at the roots of the fast solar wind, coronal hole formation, white light flares, and others will soon be discussed in detail.

Figure 4 shows jet phenomena in a polar coronal hole, as an example of coordinated observation among the *Hinode* instruments. Thanks to the broadened observable temperature range of XRT, numerous tiny loop-like structures as well as occasional ejections of X-ray jet are seen in the coronal holes. This supports changes of magnetic loop configuration as suggested by Shibata *et al.*[10] Line of sight velocities of these jets are also observed by EIS. SOT found strong horizontal magnetic structures in the coronal

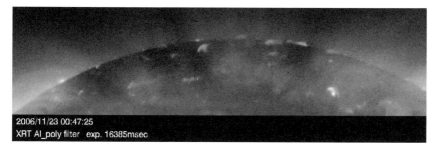

2006/11/23 00:47:25
XRT Al_poly filter   exp. 16385msec

Fig. 4.   Polar jets in the coronal hole newly found by XRT/Hinode.

holes, so that identifying the positional relationships between these strong magnetic fields and X-ray jets in the coronal holes will be a high priority task. The occurrence frequency of X-ray jets and their energetics will be of prime importance in considering the source of the high velocity solar wind, and its contribution to heating the corona in coronal holes.

4. Summary

The Japanese solar mission *"Hinode"* was successfully launched in 2006, and is providing excellent data sets in the optical, EUV, and X-ray wavelengths with high spatial and spectral resolutions. It also has polarimetric capability for measuring photospheric vector magnetic fields. All of these data obtained by the *Hinode* mission are immediately available for use by researchers and the general public. The mission has already produced nearly a hundred scientific papers on the initial results during the first year of operation.

Acknowledgments

This work is partly carried out at the NAOJ Hinode Science Center, which is supported by the Grant-in-Aid for Creative Scientific Research: "The Basic Study of Space Weather Prediction" from MEXT, Japan (Head Investigator: K. Shibata), generous donations from Sun Microsystems, and NAOJ internal funding.

References

1. T. Kosugi, K. Matsuzaki, T. Sakao, T. Shimizu, Y. Sone, S. Tachikawa, T. Hashimoto, K. Minesugi, A. Ohnishi, T. Yamada, S. Tsuneta, H. Hara,

K. Ichimoto, Y. Suematsu, M. Shimojo, T. Watanabe, S. Shimada, J. M. Davis, L. D. Hill, J. K. Owens, A. M. Title, J. L. Culhane, L. K. Harra, G. A. Doschek and L. Golub, The Hinode (Solar-B) mission: An overview, *Solar Phys. Hinode* **sp 243** (2007) 3–17.

2. K. Matsuzaki, M. Shimojo, T. D. Tarbell, L. K. Harra and E. E. Deluca, Data archive of the Hinode mission, *Solar Phys. Hinode* **sp 243** (2007) 87–92.

3. J. L. Culhane, L. K. Harra, A. M. James, K. Al-Janabi, L. J. Bradley, R. A. Chaudry, K. Rees, J. A. Tandy, P. Thomas, M. C. R. Whillock, B. Winter, G. A. Doschek, C. M. Korendyke, C. M. Brown, S. Myers, J. Mariska, J. Seely, J. Lang, B. J. Kent, B. M. Shea, R. Hagood, R. Moye, H. Hara, T. Watanabe, K. Matsuzaki, T. Kosugi, V. Hansteen and O. Wikstol, The EUV imaging spectrometer for Hinode, *Solar Phys. Hinode* **sp 243** (2007) 19–61.

4. L. Golub, E. DeLuca, G. Austin, J. Bookbinder, D. Caldwell, P. Cheimets, J. Cirtain, M. Cosmo, P. Reid, A. Sette, M. Weber, T. Sakao, R. Kano, K. Shibasaki, H. Hara, S. Tsuneta, K. Kumagai, T. Tamura, M. Shimojo, J. McCracken, J. Carpenter, H. Haight, R. Siler, E. Wright, J. Tucker, H. Rutledge, M. Barbera, G. Peres and S. Varisco, The X-ray telescope (XRT) for the Hinode mission, *Solar Phys. Hinode* **sp 243** (2007) 63–86.

5. S. Tsuneta, K. Ichimoto, Y. Katsukawa, S. Nagata, M. Otsubo, T. Shimizu, Y. Suematsu, M. Nakagiri, M. Noguchi, T. Tarbell, A. Title, R. Shine, W. Rosenberg, C. Hoffmann, B. Jurcevich, G. Kushner, M. Levay, B. Lites, D. Elmore, T. Matsushita, N. Kawaguchi, H. Saito, I. Mmikami, L. D. Hill and J. K. Owens, The solar optical telescope for the Hinode mission: An overview, *Solar Phys. Hinode* **sp 249** (2008) 167–196.

6. Y. Suematsu, S. Tsuneta, K. Ichimoto, T. Shimizu, M. Otsubo, Y. Katsukawa, M. Nakagiri, M. Noguchi, T. Tamura, Y. Kato, H. Hara, M. Kubo, I. Mikami, H. Saito, T. Matsushita, N. Kawaguchi, T. Nakaoji, K. Nagae, S. Shimada, N. Takeyama and T. Yamamuro, The solar optical telescope of SOLAR-B (HINODE): The optical telescope assembly, *Solar Phys. Hinode* **sp 249** (2008) 197–220.

7. T. Shimizu, S. Nagata, S. Tsuneta, T. Tarbell, C. Edwards, R. Shine, C. Hoffmann, E. Thomas, S. Sour, R. Rehse, O. Ito. Y. Kashiwagi, M. Tabata, K. Kodeki, M. Nagase, K. Matsuzaki, K. Kobayashi, K. Ichimoto and Y. Suematsu, Image stabilization system for Hinode (Solar-B) solar optical telescope, *Solar Phys. Hinode* **sp 249** (2008) 221–232.

8. K. Ichimoto, B. Lites, D. Elmore, Y. Suematsu, S. Tsuneta, Y. Katsukawa, T. Shimizu, R. Shine, T. Tarbell, A. Title, J. Kiyohara, K. Shinoda, G. Card, A. Lecinski, K. Streander, M. Nakagiri, M. Miyashita, M. Noguchi, C. Hoffmann and T. Cruz, Polarization calibration of the solar optical telescope onboard Hinode, *Solar Phys. Hinode* **sp 249** (2008) 233–261.

9. *Publication of Astronomical Society of Japan* Special Issue (2007) Vol. 59-S3 contains 43 papers of the initial results obtained by the *Hinode* mission.

10. K. Shibata *et al.*, *Publ. Astron. Soc. Japan* **44** (1992) L173.

Advances in Geosciences
Vol. 14: Solar Terrestrial (2007)
Eds. Marc Duldig *et al.*
© World Scientific Publishing Company

THE ROLE OF RECONNECTION IN THE CME/FLARE PROCESS

B. VRŠNAK

Hvar Observatory, Faculty of Geodesy, Kačićeva 26
HR-10000 Zagreb, Croatia
bvrsnak@geof.hr

Coronal mass ejections (CMEs) are often accompanied by an energy release in the form of a two-ribbon flare. Morphology and evolution of such flares (expanding ribbons, growing loop-system, cusp structure, etc.) are often explained in terms of reconnection of the arcade magnetic field, taking place below the erupting flux-rope. A tight relationship between the CME acceleration and the flare energy release is evidenced by various statistical correlations between parameters describing CMEs and flares, as well as by the synchronization of the CME acceleration phase with the impulsive phase of the associated flare. Such a behavior indicates that the feedback relation exists between the dynamics of CME and the reconnection process. From the theoretical point of view, reconnection affects the dynamics of the eruption in several ways. First, it reduces the tension of the overlying arcade field and increases the magnetic pressure below the flux-rope, thus enhancing the CME acceleration. Furthermore, it supplies the poloidal magnetic flux to the rope, which helps sustaining the electric current in the flux-rope and prolonging the action of the driving Lorentz force to large distances. The role of the mentioned effects is investigated by employing a simple model, where the erupting structure is represented by a curved flux-rope anchored at both ends in the dense photosphere.

1. Introduction

Although coronal mass ejections (CMEs) show a broad variety of morphological, evolutionary, and kinematical characteristics, there are some "general" properties common to most events. Typically, the eruption starts with a phase of a slow rise, often seen in measurements of the associated prominence eruption. At a certain height, the slowly rising structure suddenly starts to accelerate. The leading edge of the eruption most often accelerates by several hundreds $\mathrm{m\,s^{-2}}$,[1,2] but in most impulsive events the acceleration can attain values in the order of $10\,\mathrm{km\,s^{-2}}$.[2,3]

Maximum velocities achieved range from several tens $km\,s^{-1}$, up to more than $2000\,km\,s^{-1}$.[4]

The CME take-off is frequently accompanied by appearance of the so-called two-ribbon flare.[5,6] The eruption generally starts earlier than the two-ribbon flare energy release,[7] evidencing that such flares occur as a consequence of the eruption. Since signatures of dissipative processes (heating and particle acceleration) are usually absent in early stages of the eruption, it can be presumed that the eruption is initially driven by ideal MHD processes. However, in a certain fraction of events the flare energy release starts before the eruption.[7] This could be related to various types of pre-eruptive magnetic field restructuring that leads to loss of equilibrium and the onset of eruptive instability. For example, a confined flare can result in a structure that is kink-unstable so it erupts immediately after being formed.[8] Similarly, various forms of tether-cutting processes,[9] or break-out processes,[10] could lead to the point when the structure losses equilibrium and erupts, resulting in a CME and the associated two-ribbon flare.

The two-ribbon flare appears as a consequence of fast magnetic field reconnection that takes place in the current sheet formed in the wake of CME[6,11–13] (Fig. 1). The energy released by reconnection is transported to the chromosphere by electron beams and thermal conduction, forming there two bright ribbons aligned with the magnetic inversion line. As the reconnection proceeds, the ribbons expand away from the inversion

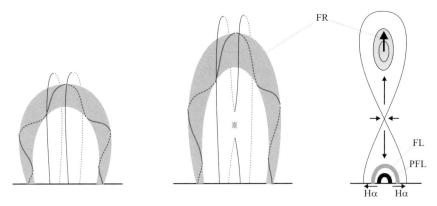

Fig. 1.   Schematic drawing of the eruption of a 3D flux-rope (left and middle); the explosion symbol indicates reconnection. In the right panel the cross section is shown (FR — flux-rope; FL — flare loops; PFL — post-flare loops; Hα — chromospheric ribbons).

line, whereas the reconnected field lines form a growing system of hot X-ray loops.[14,15]

The reconnection process beneath the erupting structure has two important effects on the CME. Firstly, the reconnection reduces the net tension of the arcade field overlying the erupting structure, which is often considered to be a magnetic flux-rope. At the same time, reconnection increases the magnetic pressure below the flux-rope, which certainly plays a significant role in the CME dynamics.[16-19] Secondly, the upward-directed reconnection jet carries the reconnected field lines to the erupting flux-rope, supplying it with a "fresh" poloidal flux. This effect enhances the hoop force,[20] thus reinforcing and prolonging the flux-rope acceleration.[21] On the other hand, the CME expansion determines the overall geometry of the system and the flows behind the flux-rope, both affecting the reconnection process.[13,22] In this way, a feedback relationship between the CME motion and the flare energy release is established. The synchronization of the CME acceleration and the energy release in the flare is most likely a consequence of such a feedback, since the reconnection rate determines also the energy release in the flare.[23-25]

The aim of this chapter is to analyze how, and to a what degree reconnection affects the CME acceleration and propagation. In Sec. 2 we present the empirical aspect of the relationship. In Secs. 3 and 4 a simple three-dimensional (3D) flux-rope model is employed to explain the observations. The results are discussed in Sec. 5.

2. Observational Evidences of Reconnection

Various observations indicate that CME dynamics is closely related to the energy release in the associated flare, or *vice versa*, the energy release in flare is tightly associated with CME kinematics. For example, statistical studies show that CME parameters, like the velocity or kinetic energy, are correlated with characteristics of the associated flare, e.g., the SXR peak flux or integrated flux.[26-28] Furthermore, case studies reveal synchronization of the flare impulsive phase and the CME acceleration stage.[29-37]

Maričić *et al.*[7] have analyzed in detail the relationship between the acceleration of the eruption and the development of energy release in the associated flare, employing a coherent sample of 22 well-observed CMEs. The main outcome of this study was that in most of events the CME acceleration phase is synchronized with the impulsive phase of the associated flare. In this way results of some previous case studies[14,31-33,36] were extended in the statistical sense, clearly demonstrating that reconnection plays a very important role in the CME acceleration.

In this respect, it is important to note that a certain fraction of ejections is not accompanied by any significant flare energy release. Nevertheless, in such events a growing system of post-eruption loops, similar to that in two-ribbon flares (post-flare loops), is often observed.[28] This implies that the magnetic field reconnection itself, rather than the flare energy release, affects the dynamics of the eruption. Indeed, measurements of the product of the flare-ribbon expansion velocity and the underlying photospheric magnetic field, representing a proxy for the coronal reconnection rate,[38] show a close synchronization of the reconnection rate and the CME acceleration.[39-41]

The active role of reconnection is documented also by morphological changes within the erupting structure. For example, in events well covered by multiwavelength observations, a hot envelope can be seen encircling the eruptive prominence.[14,34] Such a hot oval feature, detached from the solar surface, is expected to be a direct consequence of reconnection,[12,19] representing the coronal counterpart of post-flare loops (Fig. 1). A similar phenomenon can be noticed sometimes in white-light coronagraphic observations, usually denoted as disconnection events.[42-45]

The reconnection in the wake of CME can last for tens of hours. The evidence for that can be found in measurements of the growth of post-eruption loop systems.[14,46] A similar conclusion can be drawn from observations of thin streamer-like coronal features, sometimes called post-CME rays.[45,47-49]

Finally, let us mention two very interesting recent studies,[50,51] where the total flux reconnected in the CME-related flare was compared with the magnetic flux of the associated magnetic clouds measured at 1 AU. Although based on some *ad hoc* assumptions, these studies show that the poloidal flux of magnetic clouds is comparable with the total magnetic flux reconnected in the parent flare/CME process, both being significantly larger than the axial flux of the magnetic cloud. This indicates that a significant part of the magnetic flux of the interplanetary flux-rope is created by reconnection during the CME take-off.

3. 3D Flux-Rope Model without Reconnection

The magnetic field and the electric current system of the pre-eruptive coronal structure is probably very complex. However, bearing in mind that the magnetic structure is rooted in the photosphere, it is instructive to consider a very simple approximation, representing the coronal current system by a simple line-current loop, connected to the solar surface at two footpoints. Due to the high conductivity of coronal plasma and "diamagnetic" properties of the photosphere, the magnetic flux through the area encompassed by the current loop and the solar surface, Φ, can be changed only by the flux emergence or submergence, or by reconnection of coronal fields. This has very important implications on the behavior of the current system. For example, if the current-carrying loop is pushed downward, the condition $\Phi = const.$ implies that the magnetic field B between the loop and the surface has to increase. Consequently, the magnetic pressure increases, giving rise to the restoring force that tends to move the loop back toward its initial (equilibrium) position. Analogously, the restoring force acts downward if the loop is displaced upward. Thus, if the loop would be displaced, it would oscillate around the equilibrium position.[21,52] To a certain degree, the effect is analogous to the magnet levitating in a highly conductive bowl.

This effect was first recognized by Kuperus and Raadu,[53] who represented solar prominences by the horizontal line-current I located at the height h above the solar surface (for a curved flux-rope anchored at both ends in the photosphere see Ref. 54) They expressed the restoring-force effect by employing the mirror current $-I$ located at the depth h below the surface. The force exerted per unit length $(\mathrm{N\,m^{-1}})$ of an infinite straight line-current reads[53]:

$$F_{\mathrm{mc}} = \frac{\mu_0 I^2}{4\pi h}.\tag{1}$$

Hereinafter, we call this force, which is presumably caused by the induced eddy surface currents that prevent the exchange of the magnetic flux through the surface, "the mirror-current effect".

Let us now consider what would happen with the expanding current-carrying structure if the flux Φ would remain constant, and the current circuit is closed by eddy currents at the solar surface. In that case, the flux Φ is related to the current I by the expression $\Phi = LI$, where L is the inductivity of the current circuit. Since the inductivity of the current system is proportional to its size,[a] the condition $\Phi = const.$ implies that the

---

[a]Generally, the inductivity of a thin circular loop can be expressed as $L = \frac{\mu_0 C}{4\pi}[\ln\frac{\xi A}{r^2} + \frac{1}{2}]$, where C, A, and $2r$ are the circumference, area, and the diameter of the loop, respectively, whereas ξ is a constant in the order of unity.[55]

current has to decrease in the course of the eruption. The inductive decay of the current implies that the Lorentz force decreases, as well as free energy of the system ($W = LI^2/2$). The energy is converted into the kinetic and potential energies and to the work done against the "aerodynamic" drag, i.e., the energy carried away by MHD waves.[56-58]

In the next, somewhat better approximation, we can represent the erupting structure by the flux-rope. The aforementioned effects would have pretty much the same characteristics, however, now two additional effects appear, related to the curvature of the flux-rope (Fig. 2). First, the tension of axial field $B_\parallel$ gives rise to the downward force, which can be approximately expressed in the unit-length form as[52]:

$$F_t = \frac{\mu_0 I^2}{4\pi R X^2}, \tag{2}$$

where F_t is expressed in $\mathrm{N\,m^{-1}}$, $X = B_\phi/B_\parallel$ is the poloidal-to-axial field ratio at the flux-rope surface, and R is the radius of curvature of the flux-rope axis.[21] The other effect is the gradient of the poloidal field pressure ("kink-effect"). In the simplest form, the unit-length force reads[52]:

$$F_k = \frac{\mu_0 I^2}{8\pi R}, \tag{3}$$

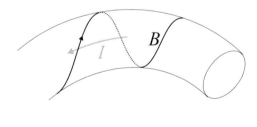

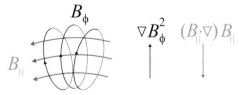

Fig. 2. A segment of curved flux-rope carrying current I. The poloidal and axial magnetic field components (B_ϕ and $B_\parallel$, respectively) are sketched. The gradient of the poloidal field pressure is directed upward, and the tension of the axial component acts downward.

which is a good approximation for the hoop force derived by applying the principle of virtual work.[59] The curvature force was treated also in some 2D spherical models[60] but the related accelerations were relatively low due to the large radius of curvature of the flux-rope encircling the sun. This force becomes effective only in 3D flux-rope models where radius of curvature is small.[54,61,62]

Unlike in 2D approach, where the flux-rope levitates within the coronal arcade,[19,63−65] in 3D models the rope is anchored in the photosphere.[20,21,52,61,66] In such situation, since the flux-rope has finite size, the overlying field can be pushed aside the eruption (Fig. 3(a)). Thus, unlike in 2D models,[11,12,19,63−65] reconnection is not necessarily the main mechanism which reduces the tension of the overlying field. In such a situation the motion through the ambient coronal field can be reproduced fairly well by employing the concept of the "aerodynamic" drag[56−58] or the snow plough effect.[67] The fast development of numerical MHD techniques in last decade enabled also detailed numerical studies of a line-tied flux-rope embedded in the coronal magnetic fields.[62,68−71]

For the matter of simplicity, let us consider a toroidal flux-rope whose axis is a circular arc anchored at fixed points in the photosphere, separated by $2d$.[21] In that case the axis length Λ and radius of curvature R are simple parametric functions of the height h of the axis summit and the footpoint

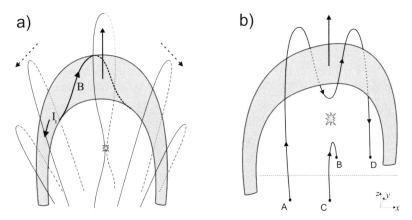

Fig. 3. (a) Schematic drawing of the eruption of a curved flux-rope anchored at both ends in the photosphere. The overlying magnetic field, internal helical field, and the electric current are indicated. The motion of the flux-rope (bold vertical arrow) pushes aside the overlying field (dashed arrows). The explosion symbol indicates reconnection below the flux-rope axis. (b) Reconnection of strongly sheared overlying field lines. The magnetic inversion line is indicated by the thin dotted line oriented in the x-direction.

half-separation d. Furthermore, we take the mass density to be uniform
within the rope, and the minor radius of the torus r to be constant along
the axis.

In the absence of reconnection or emerging flux, the poloidal flux of
the rope ("internal flux", Φ_i), as well as the flux encircled by the flux-rope
and the solar surface ("external flux", Φ_e), have to be preserved. For the
circular flux-rope axis the internal and external flux behave as $\Phi_i \propto I\Lambda$ and
$\Phi_e \propto I\Lambda[\ln \frac{8R}{r} - 2]$, respectively.[59,72] Thus, in the absence of reconnection
or emerging flux ($\Phi_i = const.$ and $\Phi_e = const.$), the current I and the
radius r have to obey:

$$I \propto \Lambda^{-1}, \tag{4}$$

and

$$r \propto R. \tag{5}$$

Equation (5) implies that the rope expands in a self-similar manner, which
is one of basic observational results.[14,34,73] The conservation of the poloidal
flux also implies that the ratio of the poloidal-to-axial field at the flux-rope
surface, $X(Z) = B_\phi/B_\|$, has to behave as[21]:

$$X \propto \frac{r}{\Lambda}. \tag{6}$$

Under the described approximations, the basic forces acting at the
flux-rope summit can be expressed in terms of the geometrical parameters
Z, R, and Λ, the flux-rope mass M, the axial electric current I, and
the poloidal-to-axial field ratio X.[21,52] Beside the three forces specified
by Eqs. (1)–(3), we consider that there is a background coronal field B_c
which protrudes through the flux-rope, giving rise to the force $F_c = IB_c$
[N m^{-1}].[52,53] Since the magnetic flux through the plane-of-axis cross section
of the flux-rope has to be constant, the background field within the rope
has to behave as $B_c \propto \Lambda^{-1}r^{-1}$. Finally, the gravitational force should be
taken into account, whereas for the matter of simplicity, we neglect the
drag term.

Bearing in mind Eq. (4) one can write $\Lambda I = \Lambda_0 I_0$, where Λ_0 and
I_0 are the flux-rope axis length and the electric current at $Z = 1$ (the
situation when $R = h = d$). In the following, the geometrical parameters
$Z = h/d$, Λ, R, and r are expressed in units of the flux-rope footpoint half-
separation d, thus $\Lambda_0 \equiv \pi$, $Z_0 = 1$, $R_0 = 1$. Superposing the mentioned

forces, the acceleration of the summit flux-rope element can be expressed in the form:

$$a = \frac{A}{\Lambda^2}\left(\frac{\Lambda}{R} - \frac{2\Lambda}{RX^2} + \frac{2\Lambda}{Z}\right) + \frac{k_1}{r\Lambda^2} - k_2 g(Z), \qquad (7)$$

where $A = \frac{\pi\mu_0 I_0^2}{4M}$. The three terms in brackets represent the forces defined by Eqs. (1)–(3). The fourth term represents the IB_c force, where the coefficient k_1 is determined by the background magnetic field strength B_c at $Z = 1$. In the gravitational term $k_2 g(Z)$, the buoyancy is taken into account through the coefficient k_2.

Equations (4)–(7) determine the dynamics of the eruption. Supplementing the drag term to Eq. (7), and bearing in mind $a = dv/dt$, Eq. (7) can be integrated to get the velocity $v(Z)$, from which $Z(t)$ is obtained.

The parameters X_0, A, k_1, and k_2 determine the height of the flux-rope axis in equilibrium (marked by black dots denoted as $Z1$ in Fig. 4(a)). Different combinations of these parameters result in stable, metastable or unstable equilibria. Stable equilibria are found for low values of X_0 and A.

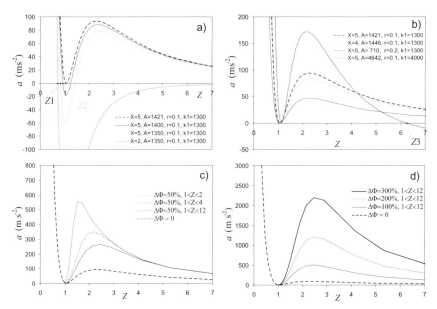

Fig. 4.   (a, b) The $a(Z)$ dependencies without reconnection; (c, d) with reconnection. The input values are written in the legend (see text for details).

Generally, for a given value of X_0, the increase of the parameter A shifts the equilibrium to larger heights. The value of the parameter A can increase either due to an increase of the flux-rope current I, or due to the mass loss. The current can increase due to emerging flux, twisting motions at the flux-rope footpoints (the poloidal flux injection analogous to that in the model by Chen and Krall[72]), the converging motions of the footpoints of the overlying arcade (like in the model by Lin et al.[12]), etc.

If X_0 is sufficiently large (generally, $X_0 \gtrsim 2$), the system evolves from the stable to a metastable state. The rope remains metastable until Z_{crit}, achieved at $A = A_{\text{crit}}$, where it eventually losses equilibrium and erupts. Such evolution is illustrated in Fig. 4(a), where the function $a(Z)$ is shown for $r_0 = 0.1$, $k_1 = -517\,\text{m s}^{-2}$ ($k_1 < 0$ means that the force is directed downward) and $k_2 = 0$ (gravitation neglected, i.e., the coronal and CME densities are equal, $\rho_c = \rho_{\text{CME}}$). The thick-gray curve represents a stable flux-rope, the black-dotted and black-full curve show two metastable situations, whereas the black-dashed curve corresponds to the flux-rope that has lost equilibrium ($a > 0$ everywhere) — with increasing value of A the equilibrium position ($Z1$ in Fig. 4(a)) rises and in the metastable state the $a < 0$ region gets successively shallower, until the dip in the curve becomes tangential with the x-axis (dashed curve in Fig. 4(a)), marking the state when the equilibrium is lost.

In Fig. 4(b) the function $a(Z)$ is shown for different values of X_0, r_0, and k_2, using $d = 0.1$ solar radii and $k_1 = 1300\,\text{m s}^{-2}$. The values of the parameter A are adjusted to the state at which the flux-rope losses equilibrium. The dashed, dotted, and full black curves are derived using $k_2 = 0$ ($\rho_c = \rho_{\text{CME}}$). Varying the value of X_0 it is found that loss of equilibrium occurs at somewhat lower heights for larger values of X_0. The thick-gray curve is obtained by using $k_2 = 1$ (negligible buoyancy, $\rho_c \ll \rho_{\text{CME}}$), showing that after a certain height the acceleration becomes negative. This means that there are situations where a stable upper equilibrium exists (denoted as $Z3$ in Fig. 4(b)).

Varying the input values over the parameter space, it is found that for physically reasonable values of X_0 and A_0, the peak accelerations do not exceed a few hundreds m s^{-2}, even assuming strong leakage of mass through the flux-rope legs.[74] In other words, the 3D flux-rope model without reconnection cannot account for the highest observed accelerations, i.e., the flux-conservation constraint prevents high accelerations because of the inductive decay of the electric current.

4. 3D Flux-Rope Model with Reconnection

The association of CMEs with two-ribbon flares and growing post-eruption loop systems evidences that reconnection is an essential ingredient of the eruption. In Fig. 3(b) the effect of the reconnection in the case of a strongly sheared arcade is sketched. The field lines initially mapping from A to B and from C to D, after reconnection connect A with D and C with B. The field line CB represents the post-eruption loop system. The AD field line is twisted around the rope, meaning that it contributes to the flux-rope current. It should be emphasized that only the x-component of the magnetic field (that is the guiding field in reconnection, since it is parallel to the photospheric magnetic inversion line) contributes to the increase of the magnetic flux Φ_e encircled by the flux-rope and the surface, since in the y-direction the reconnection by definition adds equal amount of positive and negative flux. In the case of non-sheared overlying field there is no change of the flux through the circuit, nor there is amendment of the current to the flux-rope (there is no helical field added, i.e., the added poloidal flux around the rope is purely potential). Thus, reconnection in strongly sheared arcade should support the eruption more (i.e., should provide stronger acceleration) than in weakly sheared arcade.

In the following we focus only to the effect of increasing magnetic flux encircled by the flux-rope and the solar surface Φ_e. In Figs. 4(c) and 4(d) the effect of the magnetic field reconnection, causing an increase of Φ_e, is shown. Note that, since Φ_e is not constant, the condition $r/R = const.$ is not necessarily strictly satisfied anymore. Yet, we chose the option of the self-similar expansion, which was necessary condition in the $\Phi_e = const.$ case. In such a situation the current has to behave as $I \propto \Phi_e \Lambda^{-1}$ (instead of $I \propto \Lambda^{-1}$ valid in the absence of reconnection). Due to the choice $r \propto R$, the behavior of the parameter X remains the same as in the non-reconnection case.

The curves $a(Z)$ shown in Fig. 4(c) are derived assuming that the flux increases by 50% over three different height ranges (indicated in the legend), defined by the function of the type $\Delta\Phi_e \propto \cos^2\theta$. In all cases, the values $X_0 = 5$, $r_0 = 0.1$, $k_1 = 1300\,\mathrm{m\,s^{-2}}$, and $k_2 = 0$ are applied, corresponding to the dashed curve in Figs. 4(a) and 4(b). The parameter A is adjusted to the state when the flux-rope losses equilibrium ($A = A_{\mathrm{crit}} = 1421\,\mathrm{m\,s^{-2}}$). In Fig. 4(d) we present the results for the flux change of 100%, 200%, and 300% over the distance range $1 < Z < 12$, corresponding to height range $0.1 < h < 1.2$ solar radii if $d = 0.1$. The dashed curve is the same as in Figs. 4(a) and 4(b) ($\Delta\Phi_e = 0$).

Obviously, the reconnection significantly enhances the flux-rope acceleration — now accelerations in excess of $1000\,\mathrm{m\,s^{-2}}$ can be achieved easily. Even the highest observed accelerations, in the order of $10\,\mathrm{km\,s^{-2}}$, can be attained if the flux increases by, e.g., $\approx 300\%$ over the height range $1 < Z < 2$. Bearing in mind the results by Qiu et al.[50] and Möstl et al.[51] such an increase does not contradict the observations.

With reconnection taking place over an extended period, the acceleration can be prolonged to large heights. For example, if we take $d = 0.2$ solar radii (corresponding to the footpoint separation of 280 Mm) and $\Delta\Phi_e = 100\%$ provided by reconnection taking place over the height range of $1 < Z < 12$, one finds at the radial distance of 10 solar radii ($Z = 45$) the acceleration in the order of $10\,\mathrm{m\,s^{-2}}$, consistent with the values of the Lorentz-force acceleration inferred from measurements at this height range.[58]

5. Discussion and Conclusions

Some of the basic CME characteristics can be explained already by a simple flux-rope model without reconnection. For example, it is shown that the peak acceleration and velocity, as well as the acceleration time, are determined by the initial source-region size and the magnetic field strength involved ($a \propto I_0^2/\Lambda^2$, see Eq. (7)). This explains why initially compact eruptions launched from active regions tend to accelerate more impulsively and to attain higher velocities.[2] Furthermore, the decrease of the Lorentz force with height[2,58,75] is straightforwardly explained by inductive decay of the electric current ($I \propto \Lambda^{-1}$) in the expanding structure. The model also explains the self-similar expansion of the CME.

The interplay of the mirror-current force, tension force, hoop force, and the Lorentz force associated with the background coronal field can explain existence of stable and metastable structures, as well as the gradual evolution until loss of equilibrium and eruption. Stable structures should react to external disturbances (e.g., coronal shock waves from distant flares or CMEs) by oscillating around the equilibrium, which is observed as e.g., winking filaments.[76] Similar oscillations should occur in the metastable situation if the displacement is not too large. However, if the structure is displaced up to the unstable equilibrium position (denoted in Fig. 4(a) as $Z2$), it should erupt. This explains destabilization and eruption of filaments after arrival of coronal shock waves.[77]

Generally, an increase of the electric current or increase of the magnetic flux encompassed by the current, drives the evolution of the structure from stable to metastable state. At the same time, the equilibrium height of the structure increases, which can explain the slow rising motion usually observed preceding the eruption. At a certain point, usually when the height is comparable with the footpoint half-separation, the system cannot find a neighboring equilibrium state, i.e., the equilibrium is lost and the acceleration phase starts. Model results show that the critical height is lower for more twisted structures, which is consistent with observations.[78] Furthermore, for a certain combination of the parameters describing the initial flux-rope, the model shows existence of an upper equilibrium (denoted as $Z3$ in Fig. 4(b)). Bearing in mind the damping effect of the aerodynamic drag,[79] this can explain fallback events,[80] failed eruptions,[62] and damped oscillations at the upper equilibrium.[79]

However, modeling of the flux-rope eruption in the frame of ideal MHD (no reconnection) can explain only moderate accelerations, up to a few hundreds $\mathrm{m\,s^{-2}}$. To explain the highest measured accelerations, which could be as high as $\approx 10\,\mathrm{km\,s^{-2}}$, the flux-rope model has to be extended by inclusion of reconnection, taking place beneath the rising rope. A higher reconnection rate implies a stronger acceleration, which explains good correlations between various flare and CME parameters.[26-28] The model is consistent also with the synchronization of the CME acceleration and the flare impulsive phase.[29-37] Furthermore, it explains *in situ* measurements of the magnetic fluxes of interplanetary magnetic clouds, indicating that significant part of the flux is added to the rope in the course of the eruption.[50,51]

Acknowledgments

The author would like to express his gratitude to the SOC of AOGS 2007 who kindly offered him the possibility to present invited talk at the Bangkok conference. The discussions at team meetings held in the frame of the project "Current Sheet Observations vs. Current Sheet Models" (P. I. Dr. Giannina Poletto), organized and sponsored by the International Space Science Institute (ISSI) in Bern, helped him very much in profiling and preparing this work.

References

1. J. Zhang, A study on the acceleration of coronal mass ejections, in *Coronal and Stellar Mass Ejections*, eds. K. Dere, J. Wang and Y. Yan, IAU Symposium, Vol. 226 (Cambridge University Press, 2005), p. 65.

2. B. Vršnak, D. Maričić, A. L. Stanger, A. M. Veronig, M. Temmer and D. Roša, *Solar Phys.* **241** (2007) 85.
3. D. R. Williams, T. Török, P. Démoulin, L. van Driel-Gesztelyi and B. Kliem, *Astrophys. J. Lett.* **628** (2005) L163.
4. S. Yashiro, N. Gopalswamy, G. Michalek, O. C. St. Cyr, S. P. Plunkett, N. B. Rich and R. A. Howard, *J. Geophys. Res.* (*Space Phys.*) **109** (2004) 7105.
5. B. C. Low, *Solar Phys.* **167** (1996) 217.
6. T. G. Forbes, *J. Geophys. Res.* **105** (2000) 23153.
7. D. Maričić, B. Vršnak, A. L. Stanger, A. M. Veronig, M. Temmer and D. Roša, *Solar Phys.* **241** (2007) 99.
8. H. Aurass, B. Vršnak, A. Hofmann and V. Rudžjak, *Solar Phys.* **190** (1999) 267.
9. R. L. Moore, A. C. Sterling, H. S. Hudson and J. R. Lemen, *Astrophys. J.* **552** (2001) 833.
10. S. K. Antiochos, *Astrophys. J. Lett.* **502** (1998) L181.
11. J. Lin and T. G. Forbes, *J. Geophys. Res.* **105** (2000) 2375.
12. J. Lin, J. C. Raymond and A. A. van Ballegooijen, *Astrophys. J.* **602** (2004) 422.
13. B. Vršnak and M. Skender, *Solar Phys.* **226** (2005) 97.
14. B. Vršnak, D. Maričić, A. L. Stanger and A. Veronig, *Solar Phys.* **225** (2004) 355.
15. A. M. Veronig, M. Karlický, B. Vršnak, M. Temmer, J. Magdalenić, B. R. Dennis, W. Otruba and W. Pötzi, *Astron. Astrophys.* **446** (2006) 675.
16. W. van Tend and M. Kuperus, *Solar Phys.* **59** (1978) 115.
17. U. Anzer and G. W. Pneuman, *Solar Phys.* **79** (1982) 129.
18. T. G. Forbes, *J. Geophys. Res.* **95** (1990) 11919.
19. J. Lin, *Solar Phys.* **219** (2004) 169.
20. J. Chen, *Astrophys. J.* **338** (1989) 453.
21. B. Vršnak, *Solar Phys.* **129** (1990) 295.
22. K. Shibata and S. Tanuma, *Earth, Planets, and Space* **53** (2001) 473.
23. A. Asai, T. Yokoyama, M. Shimojo, S. Masuda, H. Kurokawa and K. Shibata, *Astrophys. J.* **611** (2004) 557.
24. C. H. Miklenic, A. M. Veronig, B. Vršnak and A. Hanslmeier, *Astron. Astrophys.* **461** (2007) 697.
25. M. Temmer, A. M. Veronig, B. Vršnak and C. Miklenic, *Astrophys. J.* **654** (2007) 665.
26. Y.-J. Moon, G. S. Choe, H. Wang, Y. D. Park and C. Z. Cheng, *J. Korean Astronom. Soc.* **36** (2003) 61.
27. J. T. Burkepile, A. J. Hundhausen, A. L. Stanger, O. C. St. Cyr and J. A. Seiden, *J. Geophys. Res.* **109** (2004) A03103.
28. B. Vršnak, D. Sudar and D. Ruždjak, *Astron. Astrophys.* **435** (2005) 1149.
29. S. W. Kahler, R. L. Moore, S. R. Kane and H. Zirin, *Astrophys. J.* **328** (1988) 824.
30. W. M. Neupert, B. J. Thompson, J. B. Gurman and S. P. Plunkett, *J. Geophys. Res.* **106** (2001) 25215.
31. J. Zhang, K. P. Dere, R. A. Howard, M. R. Kundu and S. M. White, *Astrophys. J.* **559** (2001) 452.

32. P. T. Gallagher, G. R. Lawrence and B. R. Dennis, *Astrophys. J. Lett.* **588** (2003) L53.

33. A. Shanmugaraju, Y.-J. Moon, M. Dryer and S. Umapathy, *Solar Phys.* **215** (2003) 185.

34. D. Maričić, B. Vršnak, A. L. Stanger and A. Veronig, *Solar Phys.* **225** (2004) 337.

35. J. Zhang, K. P. Dere, R. A. Howard and A. Vourlidas, *Astrophys. J.* **604** (2004) 420.

36. J. Zhang and K. P. Dere, *Astrophys. J.* **649** (2006) 1100.

37. M. Temmer, A. M. Veronig, B. Vršnak, J. Rybák, P. Gömöry, S. Stoiser and D. Maričić, *Astrophys. J. Lett.* **673** (2007) 95.

38. G. Poletto and R. A. Kopp, Macroscopic electric fields during two-ribbon flares., in *The Lower Atmosphere of Solar Flares* (1986), pp. 453–465.

39. H. Wang, J. Qiu, J. Jing and H. Zhang, *Astrophys. J.* **593** (2003) 564.

40. J. Qiu, H. Wang, C. Z. Cheng and D. E. Gary, *Astrophys. J.* **604** (2004) 900.

41. J. Jing, J. Qiu, J. Lin, M. Qu, Y. Xu and H. Wang, *Astrophys. J.* **620** (2005) 1085.

42. D. F. Webb and E. W. Cliver, *J. Geophys. Res.* **100** (1995) 5853.

43. G. M. Simnett, S. J. Tappin, S. P. Plunkett, D. K. Bedford, C. J. Eyles, O. C. St. Cyr, R. A. Howard, G. E. Brueckner, D. J. Michels, J. D. Moses, D. Socker, K. P. Dere, C. M. Korendyke, S. E. Paswaters, D. Wang, R. Schwenn, P. Lamy, A. Llebaria and M. V. Bout, *Solar Phys.* **175** (1997) 685.

44. Y.-M. Wang, N. R. Sheeley, R. A. Howard, O. C. St. Cyr and G. M. Simnett, *Geophys. Res. Lett.* **26** (1999) 1203.

45. D. F. Webb, J. Burkepile, T. G. Forbes and P. Riley, *J. Geophys. Res. (Space Phys.)* **108** (2003) 1440.

46. B. Vršnak, M. Temmer, A. Veronig, M. Karlický and J. Lin, *Solar Phys.* **234** (2006) 273.

47. Y.-K. Ko, J. C. Raymond, J. Lin, G. Lawrence, J. Li and A. Fludra, *Astrophys. J.* **594** (2003) 1068.

48. J. Lin, Y.-K. Ko, L. Sui, J. C. Raymond, G. A. Stenborg, Y. Jiang, S. Zhao and S. Mancuso, *Astrophys. J.* **622** (2005) 1251.

49. A. Bemporad, G. Poletto, S. T. Suess, Y.-K. Ko, N. A. Schwadron, H. A. Elliott and J. C. Raymond, *Astrophys. J.* **638** (2006) 1110.

50. J. Qiu, Q. Hu, T. A. Howard and V. B. Yurchyshyn, *Astrophys. J.* **659** (2007) 758.

51. C. Möstl, C. Miklenic, C. J. Farrugia, M. Temmer, A. Veronig, A. B. Galvin, B. Vršnak and H. K. Biernat, *Ann. Geophys.* **26** (2008) 3139.

52. B. Vršnak, *Solar Phys.* **94** (1984) 289.

53. M. Kuperus and M. A. Raadu, *Astron. Astrophys.* **31** (1974) 189.

54. P. A. Isenberg and T. G. Forbes, *Astrophys. J.* **670** (2007) 1453.

55. J. D. Jackson, *Classical Electrodynamics*, 3rd edn. (Wiley-VCH, 1998).

56. P. J. Cargill, J. Chen, D. S. Spicer and S. T. Zalesak, *J. Geophys. Res.* **101** (1996) 4855.

57. P. J. Cargill, *Solar Phys.* **221** (2004) 135.

58. B. Vršnak, D. Ruždjak, D. Sudar and N. Gopalswamy, *Astron. Astrophys.* **423** (2004) 717.

59. T. Žic, B. Vršnak and M. Skender, *J. Plasma Phys.* **73** (2007) 741.
60. J. Lin, T. G. Forbes, P. A. Isenberg and P. Demoulin, *Astrophys. J.* **504**, (1998) 1006.
61. V. S. Titov and P. Démoulin, *Astron. Astrophys.* **351** (1999) 707.
62. T. Török and B. Kliem, *Astrophys. J. Lett.* **630** (2005) L97.
63. S. T. Wu, M. D. Andrews and S. P. Plunkett, *Space Sci. Rev.* **95** (2001) 191.
64. T. G. Forbes, J. A. Linker, J. Chen, C. Cid, J. Kóta, M. A. Lee, G. Mann, Z. Mikić, M. S. Potgieter, J. M. Schmidt, G. L. Siscoe, R. Vainio, S. K. Antiochos and P. Riley, *Space Sci. Rev.* **123** (2006) 251.
65. Z. Mikić and M. A. Lee, *Space Sci. Rev.* **123** (2006) 57.
66. T. C. Mouschovias and A. I. Poland, *Astrophys. J.* **220** (1978) 675.
67. S. J. Tappin, *Solar Phys.* **233** (2006) 233.
68. T. Amari, J. F. Luciani, J. J. Aly, Z. Mikic and J. Linker, *Astrophys. J.* **585** (2003) 1073.
69. I. I. Roussev, T. I. Gombosi, I. V. Sokolov, M. Velli, W. Manchester, IV, D. L. DeZeeuw, P. Liewer, G. Tóth and J. Luhmann, *Astrophys. J. Lett.* **595** (2003) L57.
70. B. Kliem, V. S. Titov and T. Török, *Astron. Astrophys.* **413** (2004) L23.
71. S. E. Gibson, Y. Fan, T. Török and B. Kliem, *Space Sci. Rev.* **124** (2006) 131.
72. J. Chen and J. Krall, *J. Geophys. Res.* **108** (2003) 1410.
73. V. Bothmer and R. Schwenn, *Ann. Geophys.* **16** (1998) 1.
74. B. Vršnak, V. Ruzdjak, B. Rompolt, D. Rosa and P. Zlobec, *Solar Phys.* **146** (1993) 147.
75. B. Vršnak, *J. Geophys. Res.* **106** (2001) 25249.
76. H. E. Ramsey and S. F. Smith, *Astron. J.* **71** (1966) 197.
77. H. W. Dodson, *Astrophys. J.* **110** (1949) 382.
78. B. Vršnak, V. Ruždjak and B. Rompolt, *Solar Phys.* **136** (1991) 151.
79. B. Vršnak, V. Ruždjak, R. Brajša and F. Zloch, *Solar Phys.* **127** (1990) 119.
80. Y.-M. Wang and N. R. Sheeley, Jr., *Astrophys. J.* **567** (2002) 1211.

Advances in Geosciences
Vol. 14: Solar Terrestrial (2007)
Eds. Marc Duldig *et al.*
© World Scientific Publishing Company

SMALL-SCALE INTERPLANETARY MAGNETIC FLUX ROPES AND MICRO-CMEs

D. J. WU

Purple Mountain Observatory
CAS, Nanjing, 210008, China
djwu@pmo.ac.cn

It is widely believed that interplanetary magnetic clouds (IMCs), which are large-scale interplanetary magnetic flux ropes (IMFRs) with a duration >10 h and a diameter >0.1 AU, are the manifestation of coronal mass ejections (CMEs) propagating in the interplanetary space, that is, a subset of ICMEs (interplanetary CMEs). However, recent solar wind observations show that IMFRs have a continuous distribution in size from small-scale flux ropes with a duration of tens of minutes and a diameter less than 0.01 AU to large-scale IMCs. In particular, the energy distribution of IMFRs presents a consistent power-law spectrum with a spectrum index of −1.37, which is similar to that of solar flares but with a slightly lower index. We propose that, like IMCs, the small-scale IMFRs are probably the interplanetary manifestation of micro-CMEs, which are produced in small solar eruptions, although they are too weak to appear clearly in the coronagraph observations as ordinary CMEs.

1. Introduction

The term "interplanetary magnetic cloud" (IMC) was introduced by Burlaga *et al.*[1] to describe a kind of interplanetary objects with the following properties: (1) the magnetic field strength is higher than average; (2) the magnetic field direction rotates smoothly through a large angle during an interval of the order of 1 day; and (3) the proton temperature is lower than average.[2] The two magnetic properties of IMCs suggested that IMCs are force-free magnetic field configurations, in which the magnetic field lines form a family of helices with a flux-rope geometry.[3] The magnetic field on the symmetry axis at the center of the rope is a straight line, and the pitch angle of the other field lines increases with increasing distance of the filed lines from the axis, reaching the asymptotic form of circles on the outer boundary of an IMC. Although it is very difficult to determine the global magnetic topology of an IMC from *in situ* satellite observations, its local magnetic structure can be fitted approximatively by the cylindrically symmetric constant-α force-free field with the so-called Lundquist solution.[4,5] The fitting result shows that the IMCs, as large-scale interplanetary magnetic flux ropes (IMFRs), have typical diameters of 0.20–0.40 AU.

As for the origin of IMCs, a number of evidences for an association between IMCs and coronal mass ejections (CMEs) driven by magnetic reconnection in the corona have been reported in literature since they were discovered,[6,7] and it has been accepted extensively that IMCs are an important subset of ICMEs (interplanetary CMEs), which are the direct manifestation of CMEs propagating through the interplanetary medium, and one of the most important drivers triggering geomagnetic activities.

Compared to large-scale IMFRs (i.e., IMCs) that have been extensively investigated, concerning their identification and origin, small-scale IMFRs have been given little attention due to the difficulty is identifying them. Twenty years after IMCs were discovered, Moldwin *et al.*[8] discovered a kind of small-scale IMFRs with a duration ∼1 h and a diameter ∼0.01 AU in observations of the IMP 8 and WIND spacecrafts. These small-scale IMFRs have a magnetic structure similar to that of IMCs, but their plasma temperatures are close to the average. Because the scale of this kind of IMFRs is much smaller than that of IMCs, Moldwin *et al.*[8] thought that this is a new kind of IMFRs different from IMCs in origin and proposed that the small-scale IMFRs have an interplanetary origin in comparison to the coronal origin of IMCs. A possible interplanetary source for small-scale IMFRs is magnetic reconnection across heliospheric current sheets.[8]

As pointed out by the authors, however, it is still doubtable whether IMFRs have a bi-modal or a continuous size-distribution, in particular, whether they have an interplanetary or a coronal origin. In a recent work, Feng *et al.*[9] surveyed the solar wind data observed by the WIND satellite during 1995–2001 and identified 144 IMFR events. They have durations ranging from less than 0.5 h to more than 40 h and a half of them (70 IMFRs) have durations less than 10 h. By use of the least-squares-fitting technique, the diameters of the IMFRs are calculated, and it is found that the diameters have a continuous size-distribution from 0.004 through 0.6 AU, and that half of the IMFRs have the diameters less than 0.1 AU.

In this chapter, we further discuss the energy distribution of the IMFRs, and our result shows that the IMFRs have a consistent power-law energy spectrum with a spectrum index ∼ − 1.37, which is similar to that of solar flares but slightly lower. We propose that, like IMCs, the small-scale IMFRs are probably the interplanetary manifestation of micro-CMEs, which are produced in small solar eruptions, although they are too weak to appear clearly in the coronagraph observations as ordinary CMEs.

2. Energy Spectrum of IMFRs

Following Feng *et al.*,[9] the energy of an IMFR (E) consists of the magnetic energy (E_B), thermal energy (E_T), and kinetic energy (E_K). According to the constant-α force-free field model for the IMFRs,[9] their magnetic energy in unit length can be calculated as follows:

$$E_B = 0.456(B_0^2/2\mu_0)\pi R_0^2, \tag{1}$$

where R_0 is the radius of the IMFR's cross section. The thermal energy per unit length is given by

$$E_T = 0.5n_0 m_p V_{th0}^2 \pi R_0^2, \tag{2}$$

where n_0 and V_{th0} are the mean proton density and thermal speed, respectively. Because IMFRs are moving in the ambient solar wind, the kinetic energy of IMFR per unit length can be estimated as

$$E_K = 0.5n_0 m_p \left(V_M - V_0\right)^2 \pi R_0^2, \tag{3}$$

where V_M and V_0 are the mean speed of the IMFR and the ambient solar wind velocity, respectively. In consequence, the total energy of the IMFR per unit length is

$$E = E_B + E_T + E_K \propto R_0^2. \tag{4}$$

The resulting energies of the IMFRs, per unit length along the IMFR axes, range from $\sim 10^6$ to $\sim 10^{12}$ J/m over ~ 6 orders of magnitude. In particular, the energy distribution of the IMFRs displays a continuous power-law spectrum with the spectrum index of -0.87, namely,[9]

$$F_{\text{obs}}(E) = F_0 E^{-0.87}. \tag{5}$$

Taking into account the geometry of IMFRs, an IMFR with a larger diameter is more probably observed by a satellite in the solar wind because its larger cross section may be encountered by the satellite more easily. In fact, the probability that the IMFR is observed by the satellite is directly proportional to the cross section of the IMFR. This implies that the observed probability of the IMFR is directly proportional to its radius R_0

and hence to the square root energy, $E^{1/2}$, according to the energy relation in Eq. (4), that is, $F_{obs}(E) \propto f_{act}(E)E^{1/2}$, where $f_{act}(E)$ is the actual energy spectrum of IMFRs. This indicates that more small-scale IMFRs with lower energies are missed in observations so that the actual IMFR energy spectrum should be modified as

$$f_{act}(E) = f_0 E^{-1.37}, \qquad (6)$$

which is similar to that of solar flares, but with a slightly lower spectrum index.

3. Micro-CMEs

Solar eruptions frequently produce both CMEs and flares, called flare-associated CMEs,[10,11] and it is believed that magnetic reconnection in solar magneto-atmospheres is common triggering source for both flares and CMEs. Observations of flares show that their distribution in energy have a power-law spectrum ranging from a low-energy micro-flares to high-energy large flares, namely,

$$N(E) = N_0 E^{-\alpha}, \qquad (7)$$

where the spectrum index $1.4 < \alpha < 2$.[12,13]

The result of the last section shows that the energy distribution of the IMFRs also displays a consistent power-law spectrum, which is similar to that of solar flares. On the other hand, a number of evidences demonstrate that the large-scale IMFRs of them (i.e., IMCs) are the manifestation of interplanetary propagation of CMEs, which directly originate in the coronal magnetic reconnection. In this manner, it is a plausible hypothesis that the IMFRs all originate directly from solar eruptions driven by the coronal magnetic reconnection. In particular, the small-scale IMFRs, similar to the IMCs as the manifestation of CMEs propagating through interplanetary medium, are probably the manifestation of some small-scale CMEs propagating through interplanetary medium, called micro-CMEs. The micro-CMEs have not been clearly observed, because they are too weak to appear clearly in the coronagraph observations as ordinary CMEs (e.g., the brightness range of LASCO C1, C2, C3 are $2 \times 10^{-5} \sim 2 \times 10^{-8}$, $2 \times 10^{-7} \sim 5 \times 10^{-10}$, $3 \times 10^{-9} \sim 1 \times 10^{-11} B_\odot$, respectively, where $B_\odot$ is the solar brightness,[14] and the micro-CMEs are probably out of the ranges).

Although there are no direct observations of micro-CMEs because they are too weak to appear clearly in current coronagraphs, there are a number of evidences for small solar eruptions, such as X-ray bright sigmoidal structures,[15] X-ray plasmoids,[16-20] and tiny jets observed recently by the Hinode.[21] These small eruptions may be triggered by ubiquitous magnetic reconnection between tiny emerging flux and pre-existing magnetic field, and micro-CMEs are the by-products of these small eruptions.

Because of their weakness presented in observations, it is very difficult to identify clearly the correlation between a small-scale IMFR and a special solar eruption. In recent works, Mandrini *et al.*[22,23] have provided some evidence for a small eruption observed on the solar disk center linking a small-scale IMFR with a duration of 4.12 h and a diameter of 0.0328 AU. This evidence includes the timing, the same orientation of the coronal loop and the IMFR relative to the ecliptic, the same magnetic field direction and magnetic helicity sign in the coronal loop and in the IMFR, and comparable magnetic flux measured in the dimming regions and in the IMFR. In particular, the pre- to post-event change of magnetic helicity in the solar corona is found to be consistent with the helicity content of the IMFR.

4. Discussion and Summary

Observations of solar flares show that their distribution in energy have a power-law spectrum ranging from $E < 10^{19}$ J for micro-flares and to $E > 10^{23}$ J for large flares and that the large flares frequently are accompanied by CMEs, called flare-associated CMEs. A subset of CMEs can be *in situ* observed as IMCs when they propagate through the interplanetary medium nearby the Earth. Our recent work shows that there is a continuous size-distribution of IMFRs from small-scale IMFRs with a diameter <0.01 AU to large-scale IMFRs with a diameter >0.1 AU, and it is the IMCs that are the large-scale IMFRs. In particular, the energy distribution of the IMFRs also displays a power-law spectrum similar to that of solar flares.

It is possible that, like large flares, micro-flares can be accompanied by small-scale CMEs, called micro-CMEs, although they are too weak to appear clearly in the coronagraph observations as ordinary CMEs. We propose that it is the small-scale IMFRs that are probably the interplanetary manifestation of the micro-CMEs when they propagate through interplanetary medium. In this way, the observation of the small-scale IMFRs can be treated as the evidence for the existence of the micro-CMEs, and the IMFR energy spectrum can be used to infer the

D. J. Wu

CME energy spectrum from micro-CMEs to large CMEs. In particular, the micro-CMEs present new challenges for the current CME models, which have been developed for large-scale magnetic configurations in the corona, and are potentially important for understanding the dynamics of solar magneto-atmospheres.

It is worth noticing, however, that the spectrum index of IMFRs (~1.37) is slightly lower than that of flares (~1.4–2). There are at least two possibilities to explain this difference. The first is that CMEs have a different index from that of flares. After all the two phenomena represent different processes although they are associated frequently by magnetic reconnection. In fact, flares are enhancement of electromagnetic radiation, and CMEs are ejections of magnetic plasma. If the two processes have different efficiencies in large and small scales, it is possible that they have different power-law spectrum indexes. The second possibility is that their propagation passing the interplanetary medium changes their energy distribution although CMEs have the same index as that of flares. This results in IMFRs, as their interplanetary manifestation, to have a different index. For instance, micro-CMEs are easily distorted as they move through the solar wind, and hence the number of small-scale IMFRs identified clearly in observations decrease. In consequence, the observed IMFRs have a lower spectrum index.

Acknowledgments

This project has been supported by NSFC under Grant Numbers 10425312 and 40574065, by NKBRSF under Grant Numbers 2006CB806302, and by CAS under Grant No. KJCX2-YW-T04.

References

1. L. F. Burlaga, E. Sittler, F. Mariani and R. Schwenn, *J. Geophys. Res.* **86** (1981) 6673.
2. L. F. Burlaga, in *Physics of the Inner Heliosphere II: Particles, Waves, and Turbulene*, eds. R. Schwenn and E. Marsch (Springer, New York, 1991), p. 1.
3. H. Goldstein, in *Solar Wind Five*, ed. M. Neugebauer (NASA Conference Publ. 2280, 1983), p. 731.
4. L. F. Burlaga, *J. Geophys. Res.* **93** (1988) 7217.
5. R. P. Lepping, L. F. Burlaga and J. A. Jones, *J. Geophys. Res.* **95** (1990) 11957.
6. L. F. Burlaga *et al.*, *Geophys. Res. Lett.* **9** (1982) 1317.

7. H. V. Cane and I. G. Richardson, *J. Geophys. Res.* **108** (2003) 1156.

8. M. B. Moldwin *et al.*, *Geophys. Res. Lett.* **27** (2000) 57.

9. H. Q. Feng, D. J. Wu and J. K. Chao, *J. Geophys. Res.* **112** (2007) A02102.

10. M. D. Andrews and R. A. Howard, *Space. Sci. Rev.* **95** (2001) 147.

11. Y. J. Moon *et al.*, *Astrophys. J.* **581** (2002) 694.

12. H. S. Hudson, *Solar Phys.* **133** (1991) 357.

13. I. J. D. Craig, *Solar Phys.* **202** (2001) 109.

14. G. E. Brueckne *et al.*, *Solar Phys.* **162** (1995) 357.

15. A. C. Sterling and H. S. Hudson, *Astrophys. J.* **491** (1997) L55.

16. S. Tsuneta, in *IAU Colloq. 141*, eds. H. Zirin, G. Ai and H. Wang (*ASP Conf. Proc. 46*, San Francisco, ASP, 1993), p. 239.

17. K. Shibata *et al.*, *Astrophys. J.* **451** (1995) L83.

18. M. Ohyama and K. Shibata, *Astrophys. J.* **499** (1998) 934.

19. Y. M. Wang *et al.*, *Astrophys. J.* **498** (1998) L165.

20. G. Einaudi *et al.*, *Astrophys. J.* **547** (2001) 1167.

21. K. Shibata *et al.*, in *The 4th AOGS Annual Conference: ST02 Session*, Bangkok, Thailand, 2007.

22. C. H. Mandrini *et al.*, *AAp* **434** (2005) 725.

23. C. H. Mandrini *et al.*, *Adv. Space Sci.* **36** (2005) 1579.

Advances in Geosciences
Vol. 14: Solar Terrestrial (2007)
Eds. Marc Duldig *et al.*
© World Scientific Publishing Company

RECONNECTION IN FLARES AND CMEs

G. POLETTO

INAF, Osservatorio Astrofisico di Arcetri
Largo Fermi, 5, Firenze, 50125, Italy
poletto@arcetri.astro.it

Magnetic reconnection is used nowadays to describe a wide variety of phenomena that occur throughout the universe, from the Earth's magnetosphere, to
the Sun, to accretion disks around black holes. However, we need to go back
to the late fifties to find the first suggestions about reconnection being at the
origin of solar flares. At those times the observational evidence for such an
interpretation was real scanty, with respect to the wealth of XUV, radio, and
particles data now available. This chapter reviews the observational evidence
of reconnection in flares and Coronal Mass Ejections (CMEs) provided over the
years by data mostly acquired by experiments onboard space missions. Starting
from the birth of ideas about the nature of the physical processes that might
fuel explosive transient events, first attempts will be briefly illustrated together
with basic concepts developed by Sweet–Parker and Petschek. On this basis,
a list of what we can expect to observe if reconnection is working in transient
events will be drawn, and we will show how observations met expectations. The
most recent advances will be described and future possibilities will be discussed
at the end of the chapter.

1. Origin of Explosive Events: The Birth of Ideas

Flares are observed in large numbers in the modern era, but written records,
up to the middle thirties of the last century, listed only 35 events. Following
the invention of the spectrohelioscope by Hale in 1924, the number of
records started increasing: the term *flares* dates back to those years, as
they had been previously referred to as *bright chromospheric eruptions*.[1]

We have to wait until the early sixties for the first attempts to interpret
the nature of flares. Although only qualitatively, Gold and Hoyle[2] put
forward very modern ideas when they invoked twisted magnetic loops of
opposite sense that meet and dissipate energy because of the annihilation
of the longitudinal field component.[2] The concept of a system that is driven
away from a force-free configuration and eventually dissipates energy is

used up to these days: see, e.g., a recent paper by Liu *et al.*[3] that invokes a configuration very similar to that of Gold and Hoyle as particularly flare-productive.

In the sixties, a major advancement in the interpretation of flares was reached when Sweet,[4] Parker,[5] and Petschek[6] presented their models for reconnection. These were the first models that allowed a direct comparison between theory and observations. In the Sweet and Parker reconnection, the diffusion region is much longer than its width: this hypothesis has strong consequences in that the typical time over which the reconnection process operates turns out to be too long to be applied to any real flare.

Let us open a brief digression to see how we can characterize the Sweet and Parker and Petschek reconnection. To this end we introduce the *reconnection rate* — which represents the amount of reconnected flux per unit time and is used to describe the reconnection process. In a solar flare, where the magnetic field has been torn open and returns to a relaxed state via reconnection, the rate of reconnection is given by $B_i \times v_i$, where B_i is the magnetic field just outside of the diffusion region and v_i is the plasma inflow speed (i.e., the inflow toward the current sheet — CS — where diffusion occurs).

However, usually the reconnection rate is given as an adimensional number M_0, that represents the ratio of the inflow speed to the outflow speed along the CS. Because the outflow speed is on the order of the Alfvén speed V_A ($V_A \propto B\rho^{-1/2}$, where B is the external field and ρ the CS density), $M_0 = v_i/V_A$. Alternatively, the adimensional reconnection rate M_0 may be written in terms of the *magnetic* Reynold number R_M, with $R_M = (V_A \times L)/\eta$, where L is the length of the diffusion region and η is the plasma resistivity. In this case, $M_0 = R_M^{-1/2}$, and because the magnetic Reynold number is very large (small η and large L in the Sweet–Parker regime, $R_M \approx 10^{14}$ for classical resistivity), the reconnection rate is too small to be representative of fast-eruptive events.

A solution to this problem was devised by Petschek[6] who suggested that the reconnection region did not have to be as long as suggested by Sweet[4] and Parker.[5] Petschek[6] considered the effect of waves and pointed out that it was not necessary for the whole plasma to go through the reconnection region, in order to be accelerated, as most of the acceleration might occur at the two standing shocks that form in the outflow regions. As a consequence, the diffusion region was much smaller than predicted before and bounded by two standing shocks, as if the diffusion region would bifurcate.

Figure 1 illustrates the Petschek reconnection mechanism. Although not generally accepted, the Petschek model offers an elegant solution to

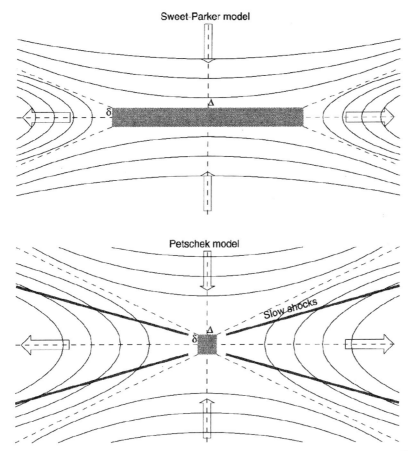

Fig. 1. Schematic representation of the steady-state Petschek reconnection. The diffusion region has a limited size: dashed lines bound the diffusion region and its "bifurcation" into two standing shocks.

the problem of the long timescales over which the Sweet and Parker model was operating, because the rate of reconnection is now proportional to $M_0 \propto \ln(R_M)^{-1/2}$, with an increase of many orders of magnitude with respect to previous values. Typically, we may say that $M_0 \approx 10^{-1} - 10^{-2}$, adequate to account for the observed phenomena. We conclude that Petschek reconnection offered to the solar physics community a plausible mechanism to interpret eruptive events.

With the advent of space era, flares and CMEs have been observed in soaring numbers. The opening of new spectral windows provided information on previously unaccessible issues, with a wealth of data that contributed to build a well-defined scenario for the process of reconnection in flares/CMEs. On one side, observations supported reconnection, and on the other, they provided new information that nourished, in a fruitful feedback, our knowledge of the process.

2. Reconnection Signatures

If we can invoke reconnection as the physical mechanism at the base of flares/CMEs, can we list, on the basis of the brief description given above, the reconnection signatures we expect to see in, for instance, flare data? This will allow us to check observations versus expectations and, possibly, devise new strategies to fill up gaps we become aware of.

Reconnection releases energy, which becomes available to reconnected fieldlines. This should lead to a *hot source* overlying or at the top of the *newly reconnected hot loops* that form as a consequence of the magnetic field reconfiguration. Their *shape* is likely to change from their primitive configuration to their relaxed state. New loops form at increasingly higher altitudes, hence, with an *increasing footpoint separation*.

We have seen how the reconnection rate depends on the plasma *inflow speed*: this, together with the *outflowing plasma jets*, is a further relevant signature. Also, we might look for an observational estimate of the *reconnection rate* as a check of consistency with predictions from Petschek model. And, whenever the X-point reconnection envisaged by the model will stretch in a *current sheet*, we will look for its properties, or, even, to its formation.

We surmise that secondary consequences of the process of reconnection should be observable when heat propagating along reconnected loops reaches down to lower chromospheric layers either as a conduction front or as accelerated particles. As a consequence of radiation/conduction cooling, cooler loops should be nested within hotter loops. These, as well as further consequences, however, will be overlooked in this paper, to keep the work as compact as possible.[a]

---

[a]The interested reader may find more complete information on topics overlooked in the present work, in, e.g., the recent book *"Reconnection of Magnetic Fields, Magnetohydrodynamics and Collisionless Theory and Observations,"* eds. J. Birn and E. R. Priest (Cambridge University Press, 2007); in the collection of papers on different aspects of CMEs in SSR, **123** (2006); and in reviews like, e.g., those of B. Vrsnak, "The role of reconnection in the CME/flare process," this volume; of J. Lin, W. Soon and S. L. Baliunas, "Theories of solar eruptions: A review," *New Astronomy Rev.* **47**, 53 (2003); of E. R. Priest and T. G. Forbes, "The magnetic nature of solar flares," *A&A Rev.* **10**, 313 (2002).

3. Signatures of Reconnection in Flares

Flares have been traditionally divided into two classes, compact and two-ribbon (or long-duration) flares. Models for two-ribbon flares have been proposed by many people, who elaborated what is well known as the CSHKP model.[7-10] In this, the magnetic field is torn open by, e.g., the eruption of a solar prominence and relaxes back to a closed configuration by reconnection of fieldlines that form increasingly higher coronal loops, while two bright chromospheric ribbons form and separate with time (see Fig. 2).

Another feature seen in two-ribbon flare as further evidence of reconnection, is the so-called loop *shrinkage*. The shrinking of loops was first suggested by Švestka *et al.*[11] who inferred this phenomenon by comparing the altitude of newly reconnected loops seen in X-rays by HXIS and FCS instruments onboard the *Solar Maximum Mission* (SMM, launched in 1980,

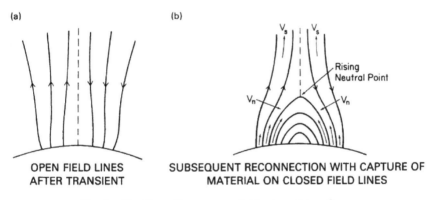

Fig. 2.   The Kopp–Pneuman model for two-ribbon flares.

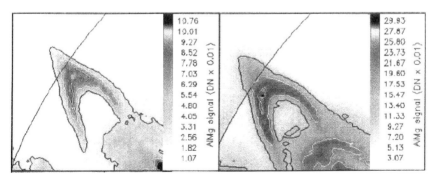

Fig. 3.   An excerpt from Ref. 13, showing the *Yohkoh* SXT signal at 21:13 UT (left) and 23:59 UT (right) for the 22–23 April 1993 flare.

see, e.g., Strong and Schmelz[12]) with the altitude of the cool H_α loops that, supposedly, were the hot loops cool remnants. Because cool loops were lower than their progenitors, the latter should *shrink*. This was an indirect evidence for a behavior characteristic of reconnection, further observed later on by, e.g., Forbes and Acton.[13] Figure 3 shows an example of shrinking loops observed by the *Yohkoh* Soft X-ray Telescope (SXT[14]) in a flare which occurred during 22–23 April 2003. Assuming that the loop shape outlines the fieldline shape, the cusp-like feature seen when loops form suggest that a neutral line lies above them, and as time goes on and new structures form, we look at the relaxation of the previously reconnected loops.

The succinct review given above shows that long-duration flares display all the reconnection signatures we may expect to observe. However, because compact flares have a smaller size, a shorter lifetime, and a lower energy release than two-ribbon flares, the mechanism that produces compact flares was not well understood until a few years ago, and the idea that a different mechanism might be invoked to explain the two-flare classes was widespread. A breakthrough observation was made by the *Yohkoh* Hard X-ray Telescope[15]: this, in conjunction with the SXT data, revealed a loop-top hard X-ray source in a compact flare, providing evidence for magnetic reconnection in these flares as well.[16]

The image at left in Fig. 4 is quite similar to the SXR image acquired by SXT. However, as we move to hardest images we see that a hot ($T_e \approx 10^8$ K) source, with no soft X-ray counterpart, shows up above soft X-ray flaring loops (the difference in altitude between the loop-tops and the hot source is $\approx 7.10^3$ km). This site has been recognized as the source of particle emission: particles originating from this location and impinging at the loop base have been invoked to account for the nearly simultaneous variation in

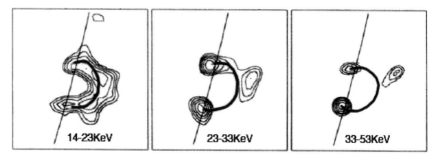

Fig. 4. An excerpt from Ref. 16, showing *Yohkoh* hard X-ray images of the 13 January 1992 flare in the three increasingly harder bands given in the panels. The field of view of each frame covers an area of 5.7×10^4 km^2. The semicircular hand-drawn thick line helps visualize the soft loops' shape in harder images (adapted from Masuda *et al.*[16]).

the footpoint emission. What has come to be known as the *Masuda flare* provided the first example hinting to a common mechanism for large and compact flares.

Continuing our checklist of the observational evidence of reconnection in eruptive events, we move on to the evidence of flows from the reconnection site. Bidirectional flows in UV observations of small-scale (order of a few arcsec), short-lived (order of a few minutes) chromospheric jets have been observed by Innes *et al.*[17] in data from the SUMER spectrometer[18] onboard the Solar Heliospheric Observatory (SOHO).[19] In a sequence of rasters, where explosive events have been seen, Doppler shifts were changing from red to blue as the spectrometer slit was scanning across the jet. The change was ascribed to the opposite direction of flows upstream of the reconnection site. Similar data have been acquired recently by Wang *et al.*,[20] still with SUMER, during a flare observed on 16 April 2002. Spectra of the impulsive flare phase show blue and red shifts in the Fe XIX 1118.1 Å line, revealing outflow speeds between 900 and $3500 \, \mathrm{km \, s^{-1}}$, consistent with theoretical predictions. That the outflows originate from the reconnection region has been unambiguously demonstrated by combining SUMER spectra with the Ramaty High-Energy Solar Spectroscopic Imager (RHESSI, see Lin *et al.*[21]) and the Transition Region and Coronal Explorer (TRACE, see Handy *et al.*[22]) observations.

A paper by Sui and Holman[23] puts together the evidence for the changing magnetic topology and the presence of flows, to build a more complete scenario of reconnection in flares. The authors analyze data from RHESSI, in different energy bands, to construct the sequence of processes that occurred in a flare observed on 15 April 2002.

What RHESSI imaged is a bright X-ray loop with a hot emission blob at its top: the hot source was associated with the reconnection point, as usual. However, after the peak of the flare, the hot source was moving outward (Fig. 5), and this separation, coupled with the changing morphology shown in the top panel of Fig. 5, was interpreted as evidence for the collapse of the X-point to form a current sheet with Y-points at its ends. As time goes on two X-ray sources thus appear at the ends of the CS. Sui and Holman[23] speculate that, during the rise phase of the flare, the magnetic configuration is of the X-type, while, during the impulsive phase, the X-type transforms into a Y-type field with fast Petschek reconnection, possibly stretching, after the flare peak, into a Sweet–Parker-type reconnection. Sui and Holman scenario provides a small-scale model reminiscent of the large-scale CME model, as indeed suggested by the authors.

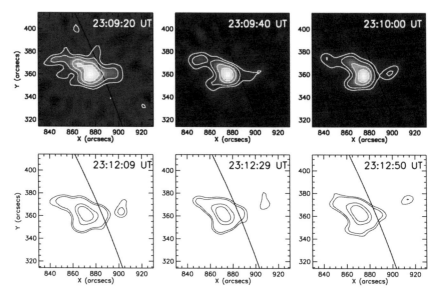

Fig. 5. A sequence of RHESSI 10–25 keV images, acquired after the peak of the
15 April 2002 flare, showing the coronal source which moves outward at an estimated
speed of 300 km s^{-1} (courtesy of Sui and Holman[23]).

4. Signatures of Reconnection in CMEs

If evidence for a CS has been found in a flare and the scenario invoked to
account for the observations is similar to the CME scenario, it should be
much easier to identify CSs in large-scale phenomena like CMEs. Figure 6
is a cartoon giving the flux rope model of Lin and Forbes.[24] The cartoon
well illustrates the relationship between the different levels involved in the
CME and looks like the Sui and Holman model expanded to larger scales.

However, as discussed by Ko *et al.*,[25] until a few years ago, there was
little observational evidence of CSs in CMEs, and this was ascribed to,
possibly, their extremely small thickness and their density, not high enough
to make the emission for the thin CS detectable. A breakthrough in this
situation came from SOHO UVCS[26] spectra that revealed high emission
from [Fe XVIII], pointing to temperatures ≥6. 10^6 K, at locations consistent
with the predicted position of CSs in CMEs.

Figure 7 illustrates the situation for the 26 November 2002, 17:00 UT
CME, analyzed by Bemporad *et al.*[27] Several of these examples have been
found in UVCS CMEs data and provided further crucial information on
the physical parameters of CSs. It turns out that CSs may have a long

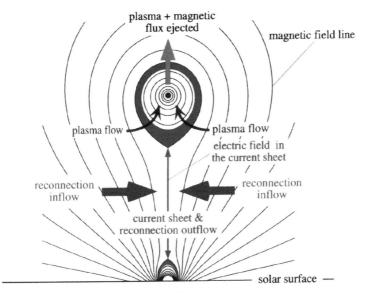

Fig. 6.   The CME model of Lin and Forbes[24] showing the CS that connects the upper part of the reconnected coronal loops to the bottom part of the CME bubble (adapted from Fig. 2 of Ko *et al.*[25]).

lifetime, over which they cool from initial temperatures above 6.10^6 K down to, eventually, the coronal temperature, have a width larger than originally thought (see, e.g., Lin *et al.*[28]), typically on the order of $1\text{–}10.10^4$ km (at heliocentric distances of the order of 1.5–2 solar radii) and a resistivity higher than electrical or anomalous resistivities. We will come back later to the implications of these results. At higher altitudes, UVCS data have been complemented by white light data: Webb *et al.*[29] studied SMM CMEs and found that bright thin structures, dubbed *rays*, are to be identified with CSs, extending more than five solar radii into the outer corona, with an average width, at 2.2 solar radii, of 7.10^4 km.

It may come as a surprise that inflows (motions toward the CS) have not been discussed, yet: in the previous section we referred to reconnection outflows, without mentioning reconnection inflows. The problem is that these are much more difficult to measure, being on the order of few $\mathrm{km\,s^{-1}}$. In flares, the first indirect evidence of inflows was presented by Yokoyama *et al.*,[30] who derived an inflow of $\approx 5\,\mathrm{km\,s^{-1}}$ from the apparent bidirectional motions of structures on either sides of a flare loop. This result was extended to more events by Narukage and Shibata[31] who give values between 2 and $15\,\mathrm{km\,s^{-1}}$, that increase to, respectively, 6 and $38\,\mathrm{km\,s^{-1}}$,

G. Poletto

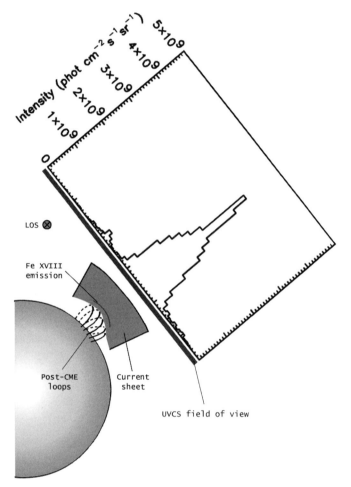

Fig. 7. A cartoon showing the position of the post-CME loops in the 26 November 2002, 17:00 UT CME and above the loops, the UVCS slit and the distribution of the [Fe XVIII] λ 974 Å line along the slit. The [Fe XVIII] emission peaks above the post-CME loops, where a CS, lying in the plane of the sky, is supposedly located.

taking into account a correction pointed out by Chen *et al.*[32] In a CME event, Lin *et al.*[33] inferred reconnection inflows between 8 and $58 \, \mathrm{km \, s^{-1}}$ with a similar indirect technique, by measuring the decrease in time of the width of a dark gap in Ly-α images. The gap was assumed to be indicative of the reconnection site, dark in Ly-α because of its high temperature. It is interesting to notice that higher inflow speeds are related to faster CMEs: an issue we will come back to later on.

5. Physical Parameters of Reconnection

The observational evidence described above leaves little doubt about the role of reconnection in flares and CMEs. Conversely, what can we infer from observations? Which information do data provide that need to be fed into reconnection theories? Here we focus on three issues, namely, what data tell us about, (a) the reconnection rate, (b) the spatial distribution of the energy release, and (c) the time-dependence of reconnection during the events.

In Sec. 1, the adimensional reconnection rate M_0 has been defined as the ratio of the inflow to the outflow speed: this definition seems to offer an easy way to derive the reconnection rate, but the problem of the detectability of inflows makes this technique hardly usable at all. A much easier way, at least for phenomena that are well described in a 2D geometry, has been suggested by Forbes and Priest.[34] Taking into account the magnetic flux conservation, the authors pointed out that the reconnection rate can be inferred from *chromospheric* data, i.e., from the speed of ribbon separation $v_{\rm rib}$ times the magnetic field at the ribbon B_n, thus avoiding uncertainties in the evaluation of coronal quantities. This equals to calculate the electric field E at the reconnection site $E = B_c \times v_c = B_n \times v_{\rm rib}$, where v_c is the inflow speed and B_c is the strength of the field at the coronal level where reconnection occurs.

The first value for the reconnection rate in a flare was derived about 20 years ago by Poletto and Kopp.[35] Over the last few years, the issue of the "observed" reconnection rate was vigorously resumed, and the increasing number of evaluations that appeared in the literature revealed a widespread interest throughout the solar community.[36-43] The derived values lie in between 0.001 and 0.18, confirming the validity of the Petschek approach.

Confirming existing theories of reconnection is not the only motivation for the interest in reconnection rates. Higher reconnection rates turn out to be associated with early flux rope acceleration in CMEs,[44] to the CME acceleration,[45] to the spectral index of the hard X-rays emitted during two-ribbon flares.[46] Hard X-rays are produced by particles accelerated by electric fields, which link them directly to the reconnection E field. Let us explore this issue in some more detail.

The ribbons we have been tacitly assuming so far, map the coronal field geometry onto the chromosphere and are observed in H_α. However, RHESSI has shown that these are accompanied by hard X-rays, which before flare maximum are seen typically as sparce footpoint sources and only after the maximum phase assume a ribbon appearance. A long-standing puzzle is to understand why the H_α and hard X-ray spatial distribution are different, if both originate from the same energy-release process. What do RHESSI *kernels*, as these concentrated hard X-ray sites are referred to, represent?

In the 2D representation assumed so far, we know that the rate of reconnection is given by $v_{\mathrm{rib}} \times B_n$, but we also have that the energy release rate is proportional to $v_{\mathrm{rib}} \times B_n^2$ (see, e.g., Isobe *et al.*[47]). In the first approximation, in an event where ribbons move more or less uniformly, we may assume that B_n and B_n^2 are proxies for the reconnection rate and the energy release rate, and we can compare the spatial distribution of the HXR intensity with that of B_n and B_n^2. Jing *et al.*,[48] analyzing data from a flare observed on 13 May 2005, found that RHESSI kernels appeared within the H_α ribbon in strong magnetic field regions, during the initial phase of the flare, but, starting from the time of flare maximum, there was no association between the HXR spatial distribution and the reconnection rate or energy. The authors surmise that this result can be explained if reconnection rearranges the magnetic field along the third direction ignored by 2D models and point out the inadequacy of the 2D representation. Most likely, whenever the flare source is a sigmoid active region (see, e.g., Sterling *et al.*,[49] Moore *et al.*[50]) reconnection can be described only with a 3D model.

Alongside the analysis of the spatial distribution of RHESSI kernels, one can analyze the temporal profiles of the reconnection and energy release rate and compare them with the HXR and microwave emission profiles (see, e.g., Kliem *et al.*,[51] Fletcher and Hudson[52]). Similarity between the profiles obviously supports magnetic reconnection as the driving flare mechanism, but also provides further information on reconnection processes. It turns out that the reconnection/energy release rate profiles, as well as the HXR and radio-emission profiles, are intermittent, revealing reconnection to be not a continuous but an intermittent phenomenon.[53] *Bursty* reconnection, as it is usually referred to, is indeed predicted by time-dependent numerical MHD simulations of the reconnection process which show the formation of magnetic islands, most likely as a result of the tearing mode instability.[54,55]

CME observations provide further evidence of intermittent reconnection. Figure 6 shows a CS connecting the system of reconnecting loops to the expanding CME bubble observed, as we said, both in UV and White Light. It is not unusual to observe *blobs* that propagate along the CS[25,33]: these have been identified with the magnetic islands predicted by numerical simulations. Furth *et al.*[56] indeed proposed more than 40 years ago that tearing instability arises whenever the CS length exceeds ≈ 6 times its width, although tearing instability is not the only mechanism invoked to account for blobs.

Observations of instabilities help us understand the CS behavior. We mentioned how the large CS thickness, observed in CMEs, came as a surprise: possibly, the formation of magnetic islands, that offer extra pressure to balance the magnetic pressure, may control the CS thickness. Lin *et al.*[57] point out in a recent work that either instabilities or Petschek time-dependent reconnection may account for the observed CS thickness and that possibly both occur in a same event. Also, CME studies, by providing measurements of the CS thickness d and estimates of the inflow speed, allowed Lin *et al.*[58] to infer values for the "observed" electrical resistivity η_e ($\eta_e \propto v_i \times d/2$) and to suggest that its unexpectedly high value may be justified if a turbulent tearing CS result in a resistivity higher than classical or anomalous resistivity, as indeed indicated by plasma turbulent theories.

6. Conclusions

After reviewing the observational evidence of reconnection in flares and CMEs, we discuss briefly the status of MHD simulations of these transient events. MHD modeling indeed represents a successful and active research area, even though there is still a gap between realistic Reynolds numbers and those used in numerical simulations. Because flares have been considered the most relevant solar transient events for a very long time, earlier numerical simulations focused on modeling the flare characteristics as they were being identified. Nowadays, eruptive flare models include all relevant flare processes like reconnection, heat conduction, chromospheric evaporation, and radiative cooling (see, e.g., Shibata[59] for a review of these works). As new data provided increasing evidence that flares and CMEs are closely related phenomena, both implying the eruption of a closed magnetic configuration, comprehensive models started to be developed.

Initial models,[60] purely hydrodynamical, modeled the response of the atmosphere to a pressure/velocity pulse applied at its base. Two-dimensional MHD processes represented the next step in these simulations[61] and have been followed by many 2D or 2.5D ideal, resistive, and hybrid MHD models, among which we may cite, those developed by, e.g., Mikić and Linker,[62] MacNeice *et al.*,[63] Van der Holst *et al.*,[64] to cite only a few of fairly numerous works. A further improvement with respect to the 2D simulations implied moving to 3D simulations.[65,66]

The progression toward 3D simulations has an important aspect, in that the topological properties of 3D fields may reveal reconnection scenarios that are not predictable in two dimensions. Among these, 3D reconnection involves the generation of electric fields with a component parallel to the magnetic field. This issue is clearly relevant to particle acceleration, opening a possibility for the interpretation of those RHESSI data (see Sec. 5) which are unaccounted for in 2D simulations.

Also, in three dimensions the two shocks of the steady Petschek reconnection (see Sec. 1) transform into shock fronts[67] that, unlike what happens in two dimensions, are not stationary, but release magnetic energy and accelerate the structure along which they propagate. Eventually, the post-flare, post/CME loop arcade may be a consequence of this behavior.

About 10 years ago, Tsuneta[68] gave a review on open issues in magnetic reconnection. It is interesting to see how many of the open issues he listed have been solved over this time interval: for instance, we described advancements made in the detection of inflows/outflows, in the CS formation, in estimating the energy provided by reconnection. Still, there are issues which need to be addressed, as, although tackled over the past decade, have not yet been conclusively solved. Among these, the following are indicated:

- what initiates reconnection? do inflows trigger reconnection or viceversa? or should we rather invoke a global instability of the magnetic configuration? and does the development of turbulence have any role in this process?
- can we trace the evolution with time of the CS width, of its length and of the reconnection rate?
- can we identify a transition between the Petschek and the Sweet–Parker regime in transient events?
- how "bursty" is bursty reconnection? do CS have a fractal structure? can we identify the spectrum of turbulence?
- is reconnection scale-invariant? is the rate of reconnection a universal constant?
- 3D reconnection is still largely to be explored: how can we infer the reconnection rate in a 3D configuration?

Space experiments like SOHO, TRACE, RHESSI have been crucial in progressing our understanding of the reconnection process in flares and CMEs over the past decade and the HINODE and STEREO missions, recently launched, have the potentiality to analogously increase our knowledge over the next few years.

Acknowledgments

The author would like to thank the SOC for giving the opportunity of presenting this talk at the ST06-10 Session of the IV AOGS 2007 Assembly. The author thanks the International Space Science Institute, ISSI, Bern for the hospitality provided to the members of the team on the Role of Current Sheets in Solar Eruptive Events, where many of the ideas presented in this work have been discussed. G. P. acknowledges support from ASI/INAF I/035/05/0 and ASI/INAF I/05/07/0.

References

1. H. W. Dodson, *Astrophys. J.* **93** (1941) 208.
2. T. Gold and F. Hoyle, *MNRAS* **120** (1960) 89.
3. Y. Liu *et al.*, *Solar Phys.* **240** (2007) 253.
4. P. A. Sweet, *Nuovo Cimento*, Suppl. **8** (1958) 188.
5. E. N. Parker, *Astrophys. J.* Suppl. **8** (1963) 177.
6. H. E. Petschek, *NASA SP50* (1964) 425.
7. H. Carmichael, *NASA SP-50* (1964) 451.
8. P. A. Sturrock, *Nature* **211** (1966) 695.
9. T. Hirayama, *Solar Phys.* **34** (1974) 323.
10. R. A. Kopp and G. W. Pneuman, *Solar Phys.* **85** (1976) 50.
11. Z. Švestka *et al.*, *Solar Phys.* **108** (1987) 237.
12. K. T. Strong and J. T. Schmelz, in *The Many Faces of the Sun*, eds. K. T. Strong, J. L. Saba, B. Haisch and J. T. Schmelz (Springer, 1999), p. 1.
13. T. G. Forbes and L. W. Acton, *Astrophys. J.* **459** (1996) 330.
14. S. Tsuneta *et al.*, *Solar Phys.* **136** (1991) 37.
15. T. Kosugi *et al.*, *Solar Phys.* **136** (1991) 17.
16. S. Masuda *et al.*, *Nature* **371** (1994) 495.
17. D. E. Innes *et al.*, *Nature* **386** (1997) 811.
18. K. Wilhelm *et al.*, *Solar Phys.* **162** (1995) 189.
19. V. Domingo *et al.*, *Solar Phys.* **162** (1995) 1.
20. T. Wang, L. Sui and J. Qiu, *Astrophys. J.* **661** (2007) L207.
21. R. P. Lin *et al.*, *Solar Phys.* **210** (2002) 3.
22. B. N. Handy *et al.*, *Solar Phys.* **187** (1999) 229.
23. L. Sui and G. D. Holman, *Astrophys. J.* **596** (2003) L251.
24. J. Lin and T. G. Forbes, *J. Geophys. Res.* **105** (2000) 2375.
25. Y.-K. Ko *et al.*, *Astrophys. J.* **594** (2003) 1068.
26. J. Kohl *et al.*, *Solar Phys.* **162** (1995) 313.
27. A. Bemporad *et al.*, *Astrophys. J.* **638** (2006) 1110.
28. J. Lin *et al.*, *Astrophys. J.* **658** (2007) L123.
29. D. F. Webb *et al.*, *J. Geophys. Res.* **108**(A12) (2003) 1440.
30. T. Yokoyama *et al.*, *Astrophys. J.* **546** (2001) L69.

31. N. Narukage and K. Shibata, *Astrophys. J.* **637** (2006) 1122.
32. P. F. Chen *et al.*, *Astrophys. J.* **602** (2004) L61.
33. J. Lin *et al.*, *Astrophys. J.* **622** (2005) 1251.
34. T. G. Forbes and E. R. Priest, in *Solar Terrestrial Physics: Present and Future*, eds. D. M. Butler and K. Papadopoulos (1984), p. 1.
35. G. Poletto and R. A. Kopp, in *The Lower Atmosphere of Solar Flares*, ed. D. F. Neidig (1986), p. 453.
36. J. Qiu *et al.*, *Astrophys. J.* (2002) 565.
37. H. Wang *et al.*, *Astrophys. J.* **593** (2003) 564.
38. L. Fletcher, J. A. Pollock and H. E. Potts, *Solar Phys.* **222** (2004) 279.
39. J. B. Noglik, R. W. Walsh and J. Ireland, *Astron. Astrophys.* **441** (2005) 353.
40. H. Isobe, H. Takasaki and K. Shibata, *Astrophys. J.* **632** (2005) 781.
41. J. L. R. Saba, T. Gaeng and T. D. Tarbell, *Astrophys. J.* **641** (2006) 1197.
42. M. Temmer *et al.*, *Astrophys. J.* **654** (2007) 665.
43. J. Lin *et al.*, *Astrophys. J.* **658** (2007) L123.
44. J. Qiu *et al.*, *Astrophys. J.* **604** (2004) 900.
45. D. Maričić *et al.*, *Solar Phys.* **241** (2007) 99.
46. C. Liu *et al.*, *Bull. Am. Astr. Soc.* **38** (2007) 214.
47. H. Isobe *et al.*, *Astrophys. J.* **566** (2002) 528.
48. J. Jing *et al.*, *Astrophys. J.* **664** (2007) L127.
49. A. C. Sterling *et al.*, *Astrophys. J.* **532** (2000) 628.
50. R. L. Moore *et al.*, *Astrophys. J.* **552** (2001) 833.
51. B. Kliem, M. Karlický and A. O. Benz, *Astron. Astrophys.* **360** (2000) 715.
52. L. Fletcher and H. Hudson, *Solar Phys.* **210** (2002) 307.
53. J. Lee, D. E. Gary and G. S. Choe, *Astrophys. J.* **647** (2006) 638.
54. P. Riley *et al.*, *Astrophys. J.* **655** (2007) 591.
55. M. Bárta, B. Vršnak and M. Karlický, *Astron. Astrophys.* **477** (2008) 649.
56. H. P. Furth, J. Kileen and M. N. Rosenbluth, *Phys. Fluids* **6** (1963) 459.
57. J. Lin, Y.-K. Ko and J. C. Raymond, *Astrophys. J.* **693** (2009) 1666.
58. J. Lin *et al.*, *Astrophys. J.* **658** (2007) L123.
59. K. Shibata, in *Proceedings of the IAU 8th Asian Pacific Regional Meeting*, ASP Conference Series, eds. S. Ikeuchi *et al.* (2000), p. 371.
60. Y. Nakagawa, S. T. Wu and E. Tandberg-Hanssen, *Solar Phys.* **41** (1975) 387.
61. R. S. Steinolfson, S. T. Wu, M. Dryer and E. Tandberg-Hanssen, *Astrophys. J.* **225** (1978) 259.
62. Z. Mikić and J. A. Linker, *Astrophys. J.* **430** (1994) 898.
63. P. MacNeice *et al.*, *Astrophys. J.* **614** (2004) 1028.
64. B. Van der Holst, C. Jacobs and S. Poedts, *Astrophys. J.* **671** (2007) L77.
65. J. Birn *et al.*, *Astrophys. J.* **541** (2000) 1078.
66. T. Amari *et al.*, *Astrophys. J.* (2003) 585.
67. M. G. Linton and D. W. Longcope, *Astrophys. J.* **642** (2006) 1177.
68. S. Tsuneta, in *Magnetic Reconnection in the Solar Atmosphere*, Vol. 111, ASP Conference Series, eds. R. D. Bentley and J. T. Mariska (1996), p. 409.

Advances in Geosciences
Vol. 14: Solar Terrestrial (2007)
Eds. Marc Duldig *et al.*
© World Scientific Publishing Company

THICKNESS AND ELECTRICAL RESISTIVITY
OF THE CURRENT SHEETS IN SOLAR ERUPTIONS

JUN LIN

National Astronomical Observatories of China/
Yunnan Astronomical Observatory, Chinese Academy of Sciences
Kunming, Yunnan 650011, P. R. China

Harvard-Smithsonian Center for Astrophysics
60 Garden Street, Cambridhe, MA 02138, USA
jlin@ynao.ac.cn
jlin@cfa.harvard.edu

Solar eruptions are created when magnetic energy is suddenly converted into heat and kinetic energies by magnetic reconnection in a field reversal region or current sheet. It is often assumed that the current sheets are too thin to be observable because the electric resistivity η_e in the sheet is taken to be very small. Theories of the tearing mode turbulence predict much faster reconnection in a thick sheet than that caused by classical and anomalous resistivities. In the present work, we show the results for the current sheet thickness d determined by analyzing a set of unique data for three eruptions observed by the UVCS and the LASCO experiments on SOHO, and deduce from these results the effective resistivity that is responsible for the rapid reconnection. We find that in these events the current sheets are observable, and have extremely large values of d and effective η_e, implying that large-scale turbulence is operating within the sheets. We also discuss the properties of the so-called hyper-resistivity caused by the tearing mode and the relation to the results of this work.

1. Introduction

Eruptive solar flares involve the formation of long current sheets connecting coronal mass ejections (CMEs) to the associated flares.[1-8] The formation of such current sheets was predicted by the catastrophe model of solar eruptions,[9,10] and has also been found in numerical experiments of CMEs using MHD codes.[11,12] Such models and simulations reproduce observed features of solar eruptions, such as the dependence of motions of flare ribbons and loops on the rate of magnetic reconnection in the current sheet,[13,14] flare-CME correlations,[15,16] and rapidly expanding CME bubbles.[17] However, little is known about the interior properties of the reconnecting current sheet during the eruption.

The high electrical conductivity and force-free environment in the corona confines the reconnection region to a thin layer, so a long and thin current sheet is expected to develop in major eruptive processes. Such features of the reconnecting current sheet allow us to treat the current sheet as an infinitely thin line when the main purpose is to study the global behavior of the eruption.[10] Although indirect evidence indicates the existence of the current sheet in the eruption (see Ref. 18 for a brief review), direct observation of a thin current sheet is difficult. This is because both size and emission of the current sheet are easily dominated by other large-scale structures nearby (e.g., see Ref. 3 for detailed discussions).

So it is often assumed that the current sheet is too thin to be observable since its thickness, d, is taken to be limited by the proton Lamor radius (Refs. 19 and 20 and references therein), which is a few tens of meters in the coronal environment. This view is based on the information about small-scale magnetic reconnection in the laboratory (with the size of meters). When the sheet forms and develops, disrupting the closed magnetic field in the solar eruption, on the other hand, the sheet could evolve and extend in length at a speed of a few times $10^2 \, \mathrm{km \, s^{-1}}$.[9 10 21] In such a highly dynamical process, the scale, especially the thickness, of the sheet may not be determined simply by the Larmor radius of particles. Instead various plasma instabilities will likely occur and play an important role in diffusing magnetic field and in governing the scale of the sheet.[22 23]

Both theories[22–26] and numerical experiments[27–29] indicate that a long current sheet can easily become unstable to the tearing mode turbulence. As the turbulence occurs, the current sheet is torn into many small scale magnetic islands or turbulence eddies (see Fig. 2 of Ref. 30). Formation of the magnetic islands broadens the current sheet.[22 23] Furthermore, the annihilation caused by the tearing mode turbulence dissipates magnetic energy in a very efficient way.[24] This implies that magnetic reconnection may be able to take place at a reasonably fast rate even if the sheet is thick.

In the next section, we present a scaling law for tearing-mode unstable current sheets, and derive a lower limit to the thickness d of the current sheet. We also analyze a set of unique data for three eruptions observed by the UVCS and the LASCO experiments on SOHO in Sec. 3, and determine both $d_{\max}$ and $d_{\min}$ for these events. In Sec. 4, we use the results obtained to estimate effective η_e, and compare them with those deduced under various conditions. Finally, Sec. 5 is a discussion and summary of this work.

2. Properties of the Current Sheets Undergoing Tearing

When the tearing mode instability develops, the scales, especially the thickness d of the current sheet and the wave number k of the turbulence are related to each other such that (e.g., see Refs. 25 and 26, pp. 177–185):

$$(\tau_A/\tau_d)^{1/4} < kl < 1, \tag{1}$$

where $l = d/2$ is the half thickness of the sheet, $\tau_A = l/V_A$ is the time that it takes to traverse the sheet at the local Alfvén speed V_A, and $\tau_d = l^2/\eta$ is the time-scale at which a typical current sheet in the conducting medium tends to diffuse, η is the magnetic diffusivity in units of $m^2 \ s^{-1}$ and is related to η_e in units of ohm m by:

$$\eta_e = \mu_0\eta, \tag{2}$$

where $\mu_0 = 4\pi \times 10^{-7} \ H \ m^{-1}$ is the magnetic permeability of free space. The tearing mode instability broadens the sheet significantly such that $\tau_d \gg \tau_A$, and the first inequality in (1) gives the lower limit of l, and the second gives the upper limit. On the other hand, inequality (1) also bound the wavelength range of the tearing mode occurring in the sheet of given thickness d.

From inequality (1), we find

$$l_{\min} = k^{-1} \left(\frac{\tau_A}{\tau_d}\right)^{1/4} = \frac{\lambda}{2\pi} \left(\frac{\tau_A}{\tau_d}\right)^{1/4},$$

where λ is the wavelength of the turbulence (see Fig. 2 of Ref. 30). According to the definitions of τ_A and τ_d, as well as the rate of magnetic reconnection,[26] we have

$$l_{\min} = \left(\frac{V_d}{V_A}\right)^{(1/4)} \frac{\lambda}{2\pi} = M_A^{1/4}\frac{\lambda}{2\pi}, \tag{3}$$

where V_d is the diffusion velocity in the sheet, and $M_A = V_d/V_A = v_i/V_A$ is the rate of magnetic reconnection measured as the diffusion speed (or the reconnection inflow speed v_i) compared to the local Alfvén speed near the sheet. Here identifying V_d with v_i is based on the fact that the plasma must carry the field lines into the sheet at the same speed as these field lines are trying to diffuse outward. Therefore, Eq. (3) relates $l_{\min}$ to M_A and λ in a simple and straightforward fashion. Observations showed that M_A ranges from 0.001 to 0.1 in real events,[3,4,31] and we shall discuss the method for determining λ from observations later on.

3. Observations and Results

Although it is not easy to observe the current sheet and magnetic reconnection process directly, more and more direct evidence of magnetic reconnection and the current sheets has accumulated during the last decade. This evidence includes fans of spike-like structures located above the flare loop system,[8,32] reconnection inflow near the current sheet,[3,4,33,46] reconnection outflows approaching to the Sun[34–37] and leaving the Sun,[3,4] as well as the high temperature plasma confined in long thin streamer-like features connecting CMEs and the associated solar flares.[2,3,6,7,38,39]

Combining the knowledge that we have collected so far allows us to look into more details of the reconnecting current sheet. For example, plasma blobs flowing along the current sheets actually constitute an important tool that may help determine the range of l according to (1). These blobs may result from either magnetic reconnection in a non-uniform plasma and magnetic field, or the tearing mode instability developing in the long thin current sheets, or other types of reconnection.[25,26,40] If the instability is their origin they are quite likely to be the magnetic islands and the distance between two successive blobs can be used to determine both $l_{\min}$ and $l_{\max}$ on the basis of (1). Below we analyze a set of unique data for three eruptions; different aspects of which were studied previously by Ciaravella et al.,[2] Ko et al.[3] and Lin et al.[4]

3.1. The 8–10 January 2002 event

This eruption started with the rapid expansion of a magnetic arcade over an active region, developed a CME, and left some thin streamer-like structures with successively growing loop systems beneath them.[3] The plasma outflow and the highly ionized plasma in these streamer-like structures, as well as the growing loops beneath them, lead to the conclusion that these structures are the current sheet produced by the eruptive process expected by theories.[9,10]

A series of plasma blobs could be seen to flow along the current sheet continuously (see Figs. 7 and 18 of Ref. 3). Five of those blobs were well observed, and allow us to measure their altitudes as functions of time. According to the images and movies from LASCO C2 and C3, we are able to find eight distances between every two successive blobs, and the smallest one is 3.1×10^5 km (see also Fig. 3 of Ref. 30). We note that there could be smaller distances present below the edge of the occulting disk of the coronagraph, but due to blocking of the disk, this might be the best one we can obtain for the time being. We use this value for λ in Eq. (3) to estimate l_{min} for this event. On the basis of analyzing the dynamical and physical properties of the current sheet, Ko *et al.*[3] found that the rate of magnetic reconnection, M_A, ranges from 0.015 to 0.03. Substituting these values into (3) gives $l_{min} = 1.7 \times 10^4$ km, and then $d_{min} = 3.4 \times 10^4$ km.

More information about d for this event can also be deduced from UVCS observations. When the current sheet was observed, the UVCS slit was located at polar angle (PA) of $78°$ and a heliocentric height of $1.53R_\odot$. Within a narrow region with $\Delta PA = 3.7°$ (the full width at the half maximum of the emission intensity distribution along the UVCS slit, see Figs. 12(k) and 12(l) of Ref. 3), strong emission from highly ionized ions (such as [Fe XVIII] $\lambda974$ and [Ca XIV] $\lambda974$) that are rarely seen in the quiet corona at this height was observed. The appearance of the [Fe XVIII] emission means a temperature near 6×10^6 K in that area, and that of [Ca XIV] means 3.2×10^6 K. Outside this region was enhanced emission from lower ionization ions such as [Fe X] $\lambda1028$ and [Si VIII] $\lambda944$. Taking the size of the high ionization region as the thickness of the sheet, we find this size equal to $3.7 \times 1.53R_\odot \times \pi/180 = 0.1R_\odot = 7.0 \times 10^4$ km. Because of projection effects and the possible complex morphology of the current sheet, this value for d is considered as the upper limit. Using the second inequality of (1) constitutes another approach to determining d_{max}, but usually gives a larger value. The range of d for this event is from 3.4×10^4 to 1.3×10^5 km, and the smaller value for d_{max} is applied.

3.2. *The 18–19 November 2003 event*

This event was well observed over the east limb by several instruments both on the ground and in space, including UVCS, LASCO, and EIT.[4] It commenced with sudden and severe stretching of a closed magnetic structure in the low corona. The two legs of the stretched structure soon started moving toward one another approaching the region where a reconnecting current sheet is presumed to lie. This region showed as a dark gap in the Lyα images of the UVCS slit, and its width decreased with time, which was ascribed to the reconnection inflow near the current sheet (see Figs. 5 and 11(a) of Ref. 4).

Decreasing of the gap width with time allows us to deduce four values of the reconnection inflow speed, v_i, at different times: 58.6, 84.6, 29.3, and 8.42 km s^{-1}. These values differ from those deduced by Lin *et al.*[4] because of improvements in measuring the gap width. The asymptotic behavior of the width-time dependence noticed by Lin *et al.*[4] implies that the reconnection inflow stopped somewhere near the edges of the sheet around 10:14 UT (see also Fig. 1 of Ref. 30). Such a tendency implies that the observed width of the gap was about 6.8×10^4 km. Again this value determines the instantaneous upper limit of d.

A group of plasma blobs were also observed to flow away from the Sun along the current sheet in the event (see Fig. 3 and the associated movies of Ref. 4, and Fig. 1 of Ref. 29). Lin *et al.*[4] identified five well observed blobs, and Riley *et al.*[29] showed four. These blobs were observed more than 6 h after the Lyα dark gap (see Fig. 5 of Ref. 4) were seen, and the distance between the two closest successive blobs is found to be about 9.1×10^5 km. Values of M_A in this events range from 0.008 to 0.18 according to the results from improved measurement. Substituting these results into (3) gives $l_{\min} = 4.3 \times 10^4$ km, and thus $d_{\min} = 8.6 \times 10^4$ km. This value is slightly larger than that deduced from the Lyα gap, which apparently constitutes an inconsistency such that $d_{\min} > d_{\max}$.

Significant broadening of the current sheet in the eruption may account for this inconsistency because the blobs were observed about 9 h later than the Lyα gap. We suggest that the tearing mode turbulence is responsible for the broadening.[23] This inconsistency might be resolved when the observations of a single typical eruption from STEREO and SOLAR-B become available.

3.3. *The 23 March 1998 event*

On 23 March 1998 UVCS observed what appears to be a striking example of the CME–current sheet–arch structure. The temperature in the current sheet determined from the ions present, in particular [Fe XVIII], is $(6\text{–}8) \times 10^6$ K, and a similar temperature is derived from the *Yohkoh* observations of the post-CME arch.[2] In this event, the high temperature emission came from a very localized and intense region in the wake of the CME (refer to

Figs. 4 and 10 of Ref. 2). The confinement of high temperature emission lines (also known as flare lines) to the small region of flare behind the CME was also observed by other instruments, CDS and SUMER, on SOHO.[38,41] The work of Ciaravella *et al.*[2] was the first to demonstrate the existence of the long current sheet developed by the eruption, as predicted by the model of Lin and Forbes.[10]

The [Fe XVIII] emission from the current sheet in this event can be easily identified from its intensity distribution along the UVCS slit, and the hot flare loops below the current sheet were also seen in the *Yohkoh*/SXT images. According to Ciaravella *et al.*,[2] the full width at half maximum of the intensity distribution yields a sheet thickness of order of 10^5 km, which is considered as d_{max}. The sheet was too faint in the LASCO images to be seen clearly, and no plasma blobs in the sheet could be recognized, so d_{min} cannot be determined for this event.

4. Electrical Resistivity of the Reconnecting Current Sheets

Magnetic reconnection that continuously dissipates the magnetic field through the current sheet requires the rate of magnetic field diffusion to balance that of the magnetic flux being brought into the sheet. This leads to (Ref. 26, p. 120)

$$v_i = \frac{\eta}{l}, \tag{4}$$

where v_i is the reconnection inflow speed in units of $\mathrm{m\,s}^{-1}$.

We note that the terms "diffusion," "dissipation," and "current sheet" used here have more general meanings than those used traditionally. They refer to any process that causes magnetic diffusion and any region where such diffusion occurs, respectively (see discussions of Lin and Forbes[10] and Forbes and Lin[9] on the relevant issues). In this sense, each parameter in (4) should be considered effective and average, and (4) is applied to dynamical processes in the present work although it was originally deduced for the steady-state reconnection.[26] This issue will be further discussed later.

So, knowing the inflow speed v_i and the sheet thickness $d = 2l$ gives the magnetic diffusivity and thus the electrical resistivity of the current sheet according to Eqs. (2) and (4). Values of v_i for the 18 November 2003 event were obtained from the decreasing gap width in the Lyα images of the UVCS slit,[4] that for the 8 January 2002 event was given by Ko *et al.*[3] and that for the 23 March 1998 event was deduced from the results of Ciaravella *et al.*[2] by assuming $M_A = 0.1$.

For comparison, the classical resistivity in the quiet corona, $\eta_c = 4\pi \times 10^2\, T^{-3/2}$ m^2 s^{-1}, and the anomalous resistivity, $\eta_a = 6.4\pi \times 10^6 n_e^{-1/2}$ m^2 s^{-1}, as the result of interactions between electrons and the low-frequency ion-acoustic turbulence, for the events studied here are evaluated as well. Here T is the plasma temperature and n_e is the density in m^{-3}, and the turbulence energy is 1% of the thermal energy (e.g., Ref. 42, pp. 80–81). Values of T and n_e are taken from Ciaravella *et al.*,[2] Raymond,[39] and Ko *et al.*,[3] and vary from 5×10^6 to 10^7 K and from 4×10^{12} to $6 \times 10^{13}\,$m^{-3}, respectively.

We notice that η_e is around 12–13 and 4–5 orders of magnitude greater than η_c and η_a, respectively. This implies that even the role of the conventional anomalous resistivity in governing the processes in the reconnecting current sheet is quite limited. Instead, formation of small scale turbulence structures, such as magnetic islands, in the current sheet plays a significant role.[27,28] Consulting works related to flare modeling in which a specific form of electrical resistivity was used,[31,43] we realize that values of η_e obtained here are even 1–2 orders of magnitude greater than those assumed for solar flares.

5. Discussions and Conclusions

Overall, for three eruptive events, we have deduced $d_{\max}$ from observations, and calculated $d_{\min}$ via combining observations and properties of the tearing mode turbulence occurring in the reconnecting current sheet. Consequently, the range of η_e for each event was estimated according to (4). This may be the first measurement of both d and η_e for solar eruptions in progress since the theory of magnetic reconnection was applied to solar flares six decades ago.[44]

Our results for d would significantly modify the present models of particle accelerations in current sheet because the values of d used in all the related works[19,20] are much smaller than what we obtained. This implies that the spectrum of energetic particles accelerated in the current sheet might not be determined by the locations they leave the sheet, which is usually suggested in those works. Instead the spectrum might be determined by other processes like turbulence.[45]

The values of η_e that we deduced for three individual events are similar in magnitude, and are all incredibly large compared to those deduced from theories of classical and anomalous resistivities, and even to those usually assumed for solar flares. This result apparently suggests a very efficient diffusion process occurring in the reconnecting current sheet. But such an unusual result is also quite probably due to using Eq. (4) to relate η to other parameters for the current sheet, which may be valid only for the diffusion caused by the classical or conventional anomalous resistivities although a justification in the effective and average sense might partly account for the result.

The diffusion process taking place in the tearing sheet (see Fig. 2(b) of Ref. 25) may not be governed by classical and anomalous resistivities at all. Instead, theories on plasma turbulence indicate that the turbulence in the tearing sheet can cause a much higher resistivity than other forms of turbulence, which is known as the hyper-resistivity D[23,24] and produce a broadening sheet.[23] Strauss (p. 416)[23] found that

$$v_i = \frac{D}{l^3} \tag{5}$$

compared to (4), where D is in units of m^4 s^{-1}. Corresponding to η_e given in (2), we calculate $D_e = \mu_0 D$ for each of the three events according to (5). We still call D_e the hyper-resistivity, and now it is in units of ohm m^3.

The value of D_e can be easily deduced for each event. This is the first such investigation for real events to our knowledge, so there is no example present that allows us to perform any comparison. But Strauss[23] found that the diffusion caused by D_e could be 10^9 times that by η_a. Further studies on properties of Eq. (5) and D_e are necessary in the future, especially because (5) was deduced in the framework of the Sweet–Parker reconnection,[23] and the results of the present work also pose a serious challenge to the existing reconnection theories that cannot rigorously handle the situations in which the size of the reconnection region is rapidly evolving.

Acknowledgments

The author's work at YNAO was supported by the Ministry of Science and Technology of China under the Program 973 grant 2006CB806303, by the National Natural Science Foundation of China under the grant 40636031 and the grant 10873030, and by the Chinese Academy of Sciences under the grant KJCX2-YW-T04 to YNAO, and he was supported by NASA grant NNX07AL72G when visiting CfA. SOHO is a joint mission of the European Space Agency and the US National Aeronautics and Space Administration.

References

1. A. Bemporad, G. Poletto, S. T. Suess, Y.-K. Ko, N. A. Schwadron, H. A. Elliot and J. C. Raymond, *Astrophys. J.* **638** (2006) 1110.
2. A. Ciaravella, J. C. Raymond, J. Li, P. Reiaer, L. D. Gardner, Y. Ko and S. Fineschi, *Astrophys. J.* **575** (2002) 1116.
3. Y. Ko, J. C. Raymond, J. Lin, G. Lawrence, J. Li and A. Fludra, *Astrophys. J.* **594** (2003) 1068.
4. J. Lin, Y. Ko, J. C. Raymond, G. A. Stenborg, Y. Jiang, S. Zhao and S. Mancuso, *Astrophys. J.* **622** (2005) 1251.
5. J. C. Raymond, A. Ciaravella, D. Dobrzycka, L. Strachan, Y. Ko, M. M. Uzzo and A.-E. Raouafi, *Astrophys. J.* **597** (2003) 1106.
6. L. Sui and G. D. Holman, *Astrophys. J.* **596** (2003) L251.
7. L. Sui, G. D. Holman and B. R. Dennis, *Astrophys. J.* **612** (2004) 546.
8. D. F. Webb, J. Burkepile, T. G. Forbes and P. Riley, *J. Geophys. Res.* **108** (2003) 1440.
9. T. G. Forbes and J. Lin, *J. Atmos. Sol.-Terr. Phys.* **62** (2000) 1499.
10. J. Lin and T. G. Forbes, *J. Geophys. Res.* **105** (2000) 2375.
11. J. A. Linker, Z. Mikić, R. Lionello, P. Riley, T. Amari and D. Odrstrcil, *Phys. Plasmas* **10** (2003) 1971.
12. P. J. MacNeice, S. K. Antiochos, A. Phillips, D. S. Spicer, C. R. DeVore and K. Olson, *Astrophys. J.* **614** (2004) 1028.
13. J. Lin, *Solar Phys.* **222** (2004) 115.
14. J. Lin, T. G. Forbes, E. R. Priest and T. N. Bungey, *Solar Phys.* **159** (1995) 275.
15. J. Lin, *Solar Phys.* **219** (2004) 169.
16. M. Zhang, L. Golub, E. DeLuca and J. Burkepile, *Astrophys. J.* **574** (2002) L97.
17. J. Lin, J. C. Raymond and A. A. van Ballegooijen, *Astrophys. J.* **602** (2004) 422.
18. T. G. Forbes and L. W. Acton, *Astrophys. J.* **459** (1996) 330.
19. Y. Litvinenko, *Astrophys. J.* **462** (1996) 997.
20. P. Wood and T. Neukirch, *Solar Phys.* **226** (2005) 73.
21. J. Lin, *Chinese J. Astron. Astrophys.* **2** (2002) 539.
22. H. R. Strauss, *Phys. Fluid* **29** (1986) 3668.
23. H. R. Strauss, *Astrophys. J.* **326** (1988) 412.
24. A. Bhattacharjee and Y. Yuan, *Astrophys. J.* **449** (1995) 739.
25. H. P. Furth, J. Killeen and M. N. Rosenbluth, *Phys. Fluids* **6** (1963) 459.
26. E. R. Priest and T. G. Forbes, *Magnetic Reconnection — MHD Theory and Applications* (Cambridge University Press, New York, 2000).
27. J. Ambrosiano, W. H. Matthaeus, M. L. Goldstein and D. Plante, *J. Geophys. Res.* **93** (1988) 14383.
28. J. F. Drake, M. Swisdak, K. M. Schoeffler, B. N. Rogers and S. Kobayashi, *Geophys. Res. Lett.* **33** (2006) L13105.
29. P. Riley, R. Lionello, Z. Mikić, J. A. Linker, E. Clark, J. Lin and Y. Ko, *Astrophys. J.* **655** (2007) 591.

30. J. Lin, J. Li, T. G. Forbes, Y.-K. Ko, J. C. Raymond and A. Vourlidas, *Astrophys. J.* **658** (2007) L123.
31. T. Yokoyama and K. Shibata, *Astrophys. J.* **436** (1994) L197.
32. Z. Švestka, F. Fárnik, H. Hudson and P. Hick, *Solar Phys.* **182** (1998) 179.
33. P. Chen, K. Shibata, D. H. Brooks and H. Isobe, *Astrophys. J.* **602** (2004) L61.
34. A. Asai, T. Yokoyama, M. Shimojo and K. Shibata, *Astrophys. J.* **605** (2004) L77.
35. D. E. McKenzie and H. S. Hudson, *Astrophys. J.* **519** (1999) L93.
36. N. R. Sheeley, Jr., H. P. Warren and Y.-M. Wang, *Astrophys. J.* **616** (2004) 1224.
37. N. R. Sheeley, Jr. and Y.-M. Wang, *Astrophys. J.* **579** (2002) 874.
38. D. E. Innes, W. Curdt, R. Schwenn, E. Solanki, G. A. Stenborg and D. E. McKenzie, *Astrophys. J.* **549** (2001) L249.
39. J. C. Raymond, in *Proceedings of the SOHO 11 Symp. on From Solar Min. to Max.: Half a Solar Cycle with SOHO* (ESA Publ. Div., Noordwijk, 2002), pp. 421–430.
40. H. K. Biernat, M. F. Heyn and V. S. Semenov, *J. Geophys. Res.* **92** (1987) 3392.
41. U. Feldman, W. Curdt, G. A. Doschek, U. Schuehle, K. Wilhelm and P. Lemaire, *Astrophys. J.* **503** (1998) 467.
42. E. R. Priest, *Solar MHD* (D. Reidel, Boston, 1982), pp. 127–129.
43. T. Miyagoshi and T. Yokoyama, *Astrophys. J.* **614** (2004) 1042.
44. R. G. Giovanelli, *Nature* **158** (1964) 81.
45. M. Onofri, H. Isliker and L. Vlahos, *Phys. Rev. Let.* **96**(15) (2006) 151102.
46. T. Yokoyama, K. Akita, T. Morimoto, K. Inoue and J. Newmark, *Astrophys. J.* **546** (2001) L69.

Advances in Geosciences
Vol. 14: Solar Terrestrial (2007)
Eds. Marc Duldig et al.
© World Scientific Publishing Company

DYNAMICS OF BEAM–PLASMA INSTABILITY AND LANGMUIR WAVE DECAY IN TWO-DIMENSIONS

L. F. ZIEBELL

Instituto de Física, UFRGS
Caixa Postal 15051
91501-970 Porto Alegre, RS, Brazil
ziebell@if.ufrgs.br

R. GAELZER

Instituto de Física e Matemática, UFPel
Caixa Postal 354, Campus UFPel
96010-900 Pelotas, RS, Brazil
rudi@ufpel.edu.br

P. H. YOON

IPST, University of Maryland, College Park, MD 20742, USA

Department of Physics, POSTECH, Pohang, Korea

yoonp@glue.umd.edu

The present chapter reports two-dimensional (2D) self-consistent solution of weak turbulence equations describing the evolution of electron–beam–plasma interaction in which quasilinear as well as nonlinear three-wave decay processes are taken into account. It is found that the 2D Langmuir wave decay processes lead to a quasi-circular ring spectrum in wave number space. Associated to this finding, there are important ramifications. First, it is seen that the 2D ring spectrum of Langmuir turbulence leads to a tendency to isotropic heating of the electrons. In the literature, the isotropization of energetic electrons, detected in the solar wind for instance, is usually attributed to pitch-angle scattering. The present finding constitutes an alternative mechanism. Second, when projected onto the one-dimensional (1D) space, the 2D ring spectrum may give a false impression of Langmuir waves inverse cascading to longer wavelength regime, when in reality, the wavelength of the turbulence does not change at all, but only the wave propagation angle changes. Although the present analysis excludes the induced scattering, which is another process potentially responsible for the inverse cascade, the present finding at least calls for an investigation into the relative efficacy of the inverse-cascading process in 1D versus 2D.

1. Introduction

The beam–plasma interaction is characterized by Langmuir turbulence and emission of electromagnetic (EM) radiation, and appears in many applications in laboratory and space plasmas. The best example of beam–plasma interaction in nature may be the solar types II and III radio bursts.[1-3] The customary explanation of these bursts is that energetic electrons in the source regions excite Langmuir turbulence (L waves) by bump-on-tail instability. Afterward, part of the wave energy in the Langmuir turbulence is converted by nonlinear mode coupling processes (decay and scattering) into EM radiation. This so-called plasma emission scenario involves generation of a back-scattered Langmuir wave, and the merging of the primary and back-scattered Langmuir waves into EM radiation at twice the plasma frequency, and/or the decay of L waves into a transverse wave at the fundamental plasma frequency and an ion-sound (S) wave. This standard scenario is much discussed in the literature, but a detailed quantitative analysis based upon actual numerical solutions of the weak turbulence equations is not available in the literature. One of the major difficulties is that the simplifying assumption of one-dimensionality (1D) cannot be made for the problem of generation of radiation. As a consequence, for a detailed numerical study of EM radiation involving decay/coalescence of L and S waves, simple two-dimensional (2D) model spectra of the primary and backward-traveling Langmuir turbulence have been assumed in the literature.[4-7]

The beam–plasma interaction problem is a natural phenomenon which is ideal for testing of various features of nonlinear plasma turbulence theories. One of these features is the so-called "Langmuir condensation" phenomenon. As explanation for this phenomenon, it is commonly assumed that the beam-generated Langmuir turbulence may suffer an inverse cascade process. As a consequence, after a long time period, long-wavelength undamped Langmuir waves can accumulate, forming the "condensation." As a mechanism to check the unlimited growth of the condensate mode, Zakharov[8] proposed the collapse of the Langmuir wave packet as the mechanism to dissipate the wave energy. The Zakharov theory of Langmuir collapse[9,10] therefore presupposes that Langmuir turbulence first undergoes inverse cascade process. However, the physics of inverse cascade of beam-generated Langmuir turbulence has been numerically studied within the context of weak turbulence theory only under simple 1D situations.[11-14] To our knowledge, the set of equations of incoherent weak turbulence theory has never been solved self-consistently in 2D. Of course, Zakharov equation itself has been solved in 2D[15] and 3D,[16] but this is not the focus of our comment.

In the solar wind, energetic electrons with energy up to 1 keV are routinely observed near 1 AU. The population of these electrons is composed of a relatively dense and cold core (~ 10 eV) and an energetic component, which is in turn usually interpreted as a combination of an isotropic halo and a field-aligned beam.[17–25] The energetic electrons are believed to be generated during solar flares, and propagate out along the open field lines. As already noted in the first paragraph, it is believed that these electrons lead to solar type III radio bursts.[1–3] The *in situ* detection of ~ 1 keV energetic (halo plus beam) electrons near 1 AU is thus highly relevant to type III bursts problem. It is well known that the initial beam electrons can undergo velocity-space plateau formation, but to account for the isotropization, it is customary to invoke pitch-angle scattering by some sort of electromagnetic fluctuation such as whistler turbulence,[26,27] or by collisional dynamics, often incorporated in the theory.[28–39]

In the present chapter we report fully self-consistent numerical solution of the set of weak turbulence equations in 2D. As a first step to the full generalization that includes EM modes, we first solve electrostatic (ES) L–S wave decay problem. In spite of this simplification, the present analysis includes higher-order nonlinearity that goes beyond the well-known quasilinear (QL) process. As already noted in the previous paragraph, full numerical solution of ES decay problem has never been obtained in 2D. Even for the simple QL relaxation problem, 2D solutions are rarely obtained in the literature. The first numerical solution of 2D QL equation for the electron–beam–plasma and bump-on-tail instability problem is given in Appert *et al.*[40] In two papers by Ishihara and Hirose,[41,42] it is solved a set of 2D QL equations involving an ion beam and an ion-acoustic instability, a problem which is similar to the problem of bump-on-tail Langmuir instability, from the numerical point of view. There are other works related to 2D numerical solutions of QL equation,[43,44] but as far as we are aware of, 2D decay problem that goes beyond QL approximation has not been solved.

As it will turn out, the findings in the present discussion will be relevant to the two above-mentioned, seemingly unrelated problems of Langmuir condensation leading to strong Langmuir turbulence modulational instability and the generation of isotropic halo distribution by beam–plasma interaction. The details will be discussed in what follows.

2. Theoretical Formulation and Numerical Setup

The wave kinetic equations for L and S waves that support QL process as well as nonlinear decay processes are given in terms of the spectral wave energy density, $I_{\mathbf{k}}^{\sigma L} = \langle E_L^{\sigma 2}(\mathbf{k}) \rangle$ and $I_{\mathbf{k}}^{\sigma S} = \langle E_S^{\sigma 2}(\mathbf{k}) \rangle$, where $E_L^{\sigma}(\mathbf{k})$ and $E_S^{\sigma}(\mathbf{k})$ represent the spectral electric field component associated with L and S waves, respectively, and where $\sigma = \pm 1$ stands for the sign of wave phase velocity. The wave kinetic equations for these waves are given by[45]

$$
\frac{\partial I_{\mathbf{k}}^{\sigma L}}{\partial t} = \frac{\pi \omega_p^2}{k^2} \int d\mathbf{v} \, \delta(\sigma \omega_{\mathbf{k}}^L - \mathbf{k} \cdot \mathbf{v})
$$

$$
\times \left(\frac{n_0 e^2}{\pi} F_e(\mathbf{v}) + \sigma \omega_{\mathbf{k}}^L \, I_{\mathbf{k}}^{\sigma L} \, \mathbf{k} \cdot \frac{\partial F_e(\mathbf{v})}{\partial \mathbf{v}} \right)
$$

$$
+ \frac{\pi e^2}{2 T_e^2} \sum_{\sigma', \sigma'' = \pm 1} \sigma \omega_{\mathbf{k}}^L \int d\mathbf{k}' \, \frac{\mu_{\mathbf{k}-\mathbf{k}'} (\mathbf{k} \cdot \mathbf{k}')^2}{k^2 k'^2 |\mathbf{k} - \mathbf{k}'|^2}
$$

$$
\times \delta(\sigma \omega_{\mathbf{k}}^L - \sigma' \omega_{\mathbf{k}'}^L - \sigma'' \omega_{\mathbf{k}-\mathbf{k}'}^S) \left[\sigma \omega_{\mathbf{k}}^L \, I_{\mathbf{k}'}^{\sigma' L} \, \frac{I_{\mathbf{k}-\mathbf{k}'}^{\sigma'' S}}{\mu_{\mathbf{k}-\mathbf{k}'}} \right.
$$

$$
\left. - \left(\sigma' \omega_{\mathbf{k}'}^L \, \frac{I_{\mathbf{k}-\mathbf{k}'}^{\sigma'' S}}{\mu_{\mathbf{k}-\mathbf{k}'}} + \sigma'' \omega_{\mathbf{k}-\mathbf{k}'}^L \, I_{\mathbf{k}'}^{\sigma' L} \right) I_{\mathbf{k}}^{\sigma L} \right], \tag{1}
$$

$$
\frac{\partial}{\partial t} \frac{I_{\mathbf{k}}^{\sigma S}}{\mu_{\mathbf{k}}} = \frac{\pi \mu_{\mathbf{k}} \omega_p^2}{k^2} \int d\mathbf{v} \, \delta(\sigma \omega_{\mathbf{k}}^S - \mathbf{k} \cdot \mathbf{v}) \left[\frac{n_0 e^2}{\pi} (F_e + F_i) \right.
$$

$$
\left. + \sigma \omega_{\mathbf{k}}^L \, \frac{I_{\mathbf{k}}^{\sigma S}}{\mu_{\mathbf{k}}} \, \mathbf{k} \cdot \frac{\partial}{\partial \mathbf{v}} \left(F_e + \frac{m_e}{m_i} F_i \right) \right]
$$

$$
+ \frac{\pi e^2}{4 T_e^2} \sum_{\sigma', \sigma'' = \pm 1} \sigma \omega_{\mathbf{k}}^L \int d\mathbf{k}' \, \frac{\mu_{\mathbf{k}} [\mathbf{k}' \cdot (\mathbf{k} - \mathbf{k}')]^2}{k^2 k'^2 |\mathbf{k} - \mathbf{k}'|^2}
$$

$$
\times \delta(\sigma \omega_{\mathbf{k}}^S - \sigma' \omega_{\mathbf{k}'}^L - \sigma'' \omega_{\mathbf{k}-\mathbf{k}'}^L) \left[\sigma \omega_{\mathbf{k}}^L \, I_{\mathbf{k}'}^{\sigma' L} \, I_{\mathbf{k}-\mathbf{k}'}^{\sigma'' L} \right.
$$

$$
\left. - \left(\sigma' \omega_{\mathbf{k}'}^L \, I_{\mathbf{k}-\mathbf{k}'}^{\sigma'' L} + \sigma'' \omega_{\mathbf{k}-\mathbf{k}'}^L \, I_{\mathbf{k}'}^{\sigma' L} \right) \frac{I_{\mathbf{k}}^{\sigma S}}{\mu_{\mathbf{k}}} \right]. \tag{2}
$$

The first term in Eq. (1), which rules the evolution of L waves, is constituted by two parts, one corresponding to the spontaneous emission effect and the other corresponding to the usual quasilinear effect. The second term contains the energy conservation condition,

$\delta(\sigma\omega_{\mathbf{k}}^L - \sigma'\omega_{\mathbf{k}'}^L - \sigma''\omega_{\mathbf{k}-\mathbf{k}'}^S)$, and describes the three-wave decay process. Similar terms occur in Eq. (2), which rules the evolution of S waves.

The electron particle kinetic equation is given by

$$\frac{\partial F_e(\mathbf{v})}{\partial t} = \frac{\partial}{\partial v_i}\left(A_i(\mathbf{v})\,F_e(\mathbf{v}) + D_{ij}(\mathbf{v})\,\frac{\partial F_e(\mathbf{v})}{\partial v_j}\right),$$

$$A_i(\mathbf{v}) = \frac{e^2}{4\pi m_e}\int d\mathbf{k}\,\frac{k_i}{k^2}\sum_{\sigma=\pm 1}\sigma\omega_{\mathbf{k}}^L\,\delta(\sigma\omega_{\mathbf{k}}^L - \mathbf{k}\cdot\mathbf{v}), \qquad (3)$$

$$D_{ij}(\mathbf{v}) = \frac{\pi e^2}{m_e^2}\int d\mathbf{k}\,\frac{k_i\,k_j}{k^2}\sum_{\sigma=\pm 1}\delta(\sigma\omega_{\mathbf{k}}^L - \mathbf{k}\cdot\mathbf{v})\,I_{\mathbf{k}}^{\sigma L}.$$

The term with coefficient A_i describes the effect of spontaneous fluctuations, and the term with coefficient D_{ij} corresponds to the usual quasilinear effect, which causes particle diffusion in velocity space. For the ions, it is reasonable to assume a time-stationary state. In 2D, the ion distribution is $F_i(\mathbf{v}) = (2\pi T_i/m_i)^{-1/2}\exp\left(-m_i v^2/2T_i\right)$, where T_i and m_i stand for ion temperature and proton mass, respectively. In the above $\omega_p = (4\pi n_0 e^2/m_e)^{1/2}$ is the electron plasma frequency; and e, m_e, and n_0 stand for the unit electric charge, electron mass, and the ambient particle number density, respectively. The dispersion relations for L and S modes are well-known: $\omega_{\mathbf{k}}^L = \omega_p\left(1 + 3k^2\lambda_D^2/2\right)$, and $\omega_{\mathbf{k}}^S = \omega_p\,k\lambda_D\,(m_e/m_i)^{1/2}$ $(1 + 3T_i/T_e)^{1/2}\,(1 + k^2\lambda_D^2)^{-1/2}$, where $v_e = (2T_e/m_e)^{1/2}$ is the electron thermal speed, and $\lambda_D = v_e/(\sqrt{2}\omega_p)$ is the electron Debye length, with T_e being the electron temperature. In Eqs. (1) and (2) we have also introduced a quantity $\mu_{\mathbf{k}} = |k|^3\lambda_D^3\,(m_e/m_i)^{1/2}\,(1 + 3T_i/T_e)^{1/2}$.

We initialize the wave intensities by balancing the spontaneous and induced emissions, taking into account the background electron populations:

$$I_{\mathbf{k}}^{\sigma L}(0) = \frac{T_e}{4\pi^2}\frac{1}{1 + 3k^2\lambda_D^2},$$

$$I_{\mathbf{k}}^{\sigma S}(0) = \frac{T_e}{4\pi^2}\,k^2\lambda_D^2\,\sqrt{\frac{1 + k^2\lambda_D^2}{1 + 3k^2\lambda_D^2}}\,\frac{\int d\mathbf{v}\,\delta(\sigma\omega_{\mathbf{k}}^S - \mathbf{k}\cdot\mathbf{v})\,(F_e + F_i)}{\int d\mathbf{v}\,\delta(\sigma\omega_{\mathbf{k}}^S - \mathbf{k}\cdot\mathbf{v})\,[F_e + (T_e/T_i)\,F_i]}. \qquad (4)$$

The initial electron distribution corresponds to a combined Maxwellian background plus a tenuous component of the drifting Gaussian beam population. In 2D, it is given by $F_e(\mathbf{v}, 0) = (1 - n_b/n_0)(\pi v_e^2)^{-1}$ $\exp(-v^2/v_e^2) + (n_b/n_0)(\pi v_b^2)^{-1} \exp[-v_\perp^2/v_b^2 - (v_\parallel - v_0)^2/v_b^2]$. Here $v_e^2 = 2T_e/m_e$ and $v_b^2 = 2T_b/m_e$. We have solved Eqs. (1)–(3) in 2D wave number and 2D velocity spaces, by employing the Runge–Kutta procedure with adjustable time step for the wave equations. The value of the time step obtained with the Runge–Kutta procedure is then used as the step in the particle kinetic equation, which is solved by implicit method in alternate direction (the so-called ADI method).

We used 51×51 grids for the dimensionless variables $q_\perp = k_\perp v_e/\omega_p$ and $q_\parallel = k_\parallel v_e/\omega_p$, with $0 < q_\perp < 0.5$, and $0 < q_\parallel < 0.5$. For the velocities, we use 51×51 grids for the dimensionless variable $u_\perp = v_\perp/v_e$ and 51×101 grids for the dimensionless velocity $u_\parallel = v_\parallel/v_e$, covering the velocity range $0 < u_\perp < 12$ and $-12 < u_\parallel < 12$. We assumed the ratio of beam-to-thermal electrons $n_b/n_0 = 4.0 \times 10^{-4}$, normalized beam speed $u_0 = v_0/v_e = 5.0$, equal beam-to-background temperature $T_b/T_e = 1.0$, and the electron-to-ion temperature ratio $T_e/T_i = 7.0$. We deliberately chose this set of parameters, which are almost the same as those used in the case of our earlier 1D solution,[12] in order to be able to make comparisons between the 1D and 2D cases. The difference is that we assume a slightly larger ratio n_b/n_0 than that used in the previous 1D analysis, in order to enhance the effects connected to the beam, and that in the present formulation the initial wave level is obtained self-consistently by means of Eq. (5), while it was assumed arbitrarily in the case of the 1D analysis of Ziebell *et al.*[12] We assume the plasma parameter, which appears in the spontaneous emission effect and which was not taken into account for the 1D analysis of Ziebell *et al.*,[12] to be $(n_0 \lambda_D^3)^{-1} = 8.0 \times 10^{-3}$.

3. Numerical Results and Analysis

In Fig. 1 we show the evolution of electron distribution function in velocity space $(u_\parallel, u_\perp)$, starting from $t = 0$ to $t = 2 \times 10^3 \omega_p^{-1}$ to $t = 5 \times 10^3 \omega_p^{-1}$. The well-known quasilinear plateau formation along parallel direction can be seen to occur. However, a novel feature inherent to 2D nonlinear dynamics emerges. Figure 1 shows that the region of the parallel plateau in velocity space is also flattened in the perpendicular direction, indicating the occurrence of perpendicular heating of the distribution function. This behavior has yet to be investigated for other values of the plasma parameters, but it indicates a tendency to isotropization of the distribution function. In the present case it is basically due to the

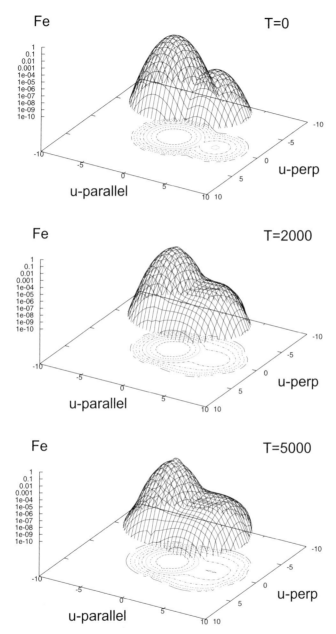

Fig. 1. Electron distribution function, $F_e(u_\perp, u_\parallel, 0)$, versus $u_\perp = v_\perp/v_e$ and $u_\parallel = v_\parallel/v_e$, in vertical logarithmic scale; (top) initial electron distribution; (middle) the electron distribution at $\omega_p t = 2 \times 10^3$; and (bottom) at $\omega_p t = 5 \times 10^3$.

quasilinear effect, but some preliminary results obtained assuming higher level of S waves and consequently more efficient three-wave decay processes show enhanced isotropization, and therefore indicates that the nonlinear evolution associated to three-wave decay can be considered as a mechanism potentially applicable to the problem of generation of isotropic solar wind electrons.

This finding is relevant to the solar wind observation of isotropic halo plus field-aligned beam distribution.[17-25] Traditional theories to account for the isotropization of the halo population rely on Coulomb collisions near the surface of the sun, and transport of the electrons through the inhomogeneous interplanetary space.[28-39] In this customary theory, collective dynamics is largely ignored, but the collisionality of the plasma near the sun, and propagation effects through largely collisionless interplanetary space are taken together. Another more recent idea invokes the pitch-angle scattering by some sort of EM fluctuation such as whistler turbulence.[26,27] In this approach, in which the presence, or excitation of EM fluctuation is the key, the isotropization process takes place locally. The present finding can be considered as yet another mechanism that can be potentially applicable for isotropic solar wind electron problem. As Fig. 1 shows, a purely electrostatic process alone is capable of naturally produce perpendicular heating. We have also obtained some preliminary results which show that with sufficiently high level of in-sound waves, quasi-isotropic energetic halo distribution can be generated, provided dynamics in the 2D space is taken into account. This isotropization mechanism also works on a local level, but unlike the whistler pitch-angle theory discussed in Refs. 26 and 27 which is essentially quasilinear, the present approach relies on nonlinear mechanism.

Figures 2 and 3 display the spectra of L and S waves for $\omega_p t = 500$, 2000, and 5000. We chose to show the starting L and S wave spectra at $\omega_p t = 500$ rather than at $t = 0$ because initially the wave intensities are rather low, and do not show any interesting features.

Figure 2 shows the appearance of a peak in the Langmuir spectrum at $\omega_p t = 500$. This time period is relatively early in the nonlinear stage, and can be viewed as the late quasilinear stage. The peak (primary) Langmuir spectrum is centered at $q_\perp = 0$ and $q_\parallel = k_\parallel v_e / \omega_p \simeq 0.2$, in accordance with linear growth theory. Note that the 2D spectrum features much broader spread along $q_\perp$ than along $q_\parallel$. At $\omega_p t = 500$ the half-width along $q_\parallel$ is nearly 0.07, and along $q_\perp$ is nearly 0.26. Another feature to be noticed is that if we were to project the results to 1D $q_\parallel$ axis, then

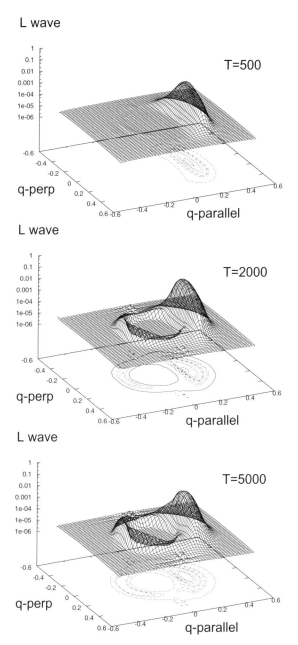

Fig. 2. *L* wave intensity, at $\omega_p t = 500$ (top), $\omega_p t = 2 \times 10^3$ (middle), and at $\omega_p t = 5 \times 10^3$, versus $q_\perp = k_\perp v_e/\omega_p$ and $q_\parallel = k_\parallel v_e/\omega_p$, in vertical logarithmic scale.

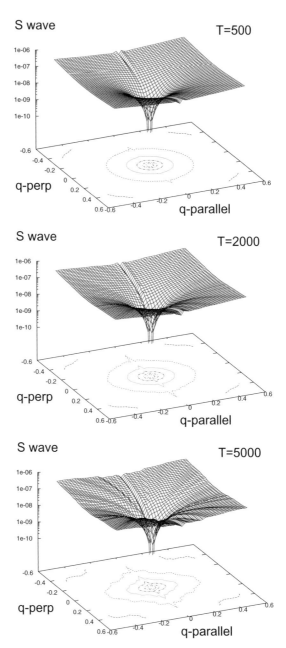

Fig. 3. S wave intensity, at $\omega_p t = 500$ (top), $\omega_p t = 2 \times 10^3$ (middle), and at $\omega_p t = 5 \times 10^3$, versus $q_\perp = k_\perp v_e / \omega_p$ and $q_\parallel = k_\parallel v_e / \omega_p$, in vertical logarithmic scale.

the primary Langmuir turbulence spectrum would be qualitatively similar to that obtained with the 1D approximation. Figure 2 also shows a slight enhancement in the backward-propagating direction with peak spectrum occurring at $q_\| = k_\| v_e / \omega_p \simeq -0.2$, which is barely visible at $\omega_p t = 500$ but is already well defined at $\omega_p t = 2.0 \times 10^3$. The asymmetric twin-peaked structure is again reminiscent of the 1D analysis.[12] However, the notable difference with the 1D case is that between the twin peaks, there is a large area in wave number space, along a circumference, where the waves are beginning to grow. This is an indication that the nonlinear decay process is already at work.

The lower panel in Fig. 2 shows a later stage of the Langmuir wave decay dynamics (at $\omega_p t = 5 \times 10^3$). While the forward-propagating peak corresponding to primary Langmuir waves was still seen to grow at $\omega_p t = 2000$, by the time that the system has evolved to $\omega_p t = 5000$ it have practically ceased to grow, indicating saturation. The backward peak, on the other hand, continues to evolve between $\omega_p t = 2000$ and $\omega_p t = 5000$. An interesting feature is that the circular ring-like structure in the spectrum of Langmuir waves, which was nearly unnoticed at $\omega_p t = 500$ and becoming visible at $\omega_p t = 2.0 \times 10^3$, continues to grow due to 2D decay processes and become a fully developed ring at $\omega_p t = 5.0 \times 10^3$. Of course, this ring-like feature could not have been predicted by simple 1D analysis.

The evolution of the S wave spectrum appears in Fig. 3. It is noticeable that the S wave turbulence spectrum develops some ripples along the time evolution but does not actually grow above the initial fluctuation level. Thus, the S mode turbulence features nothing special as far as the spectrum is concerned. However, it is important to note that even though the S waves do not get amplified, they nevertheless participate in the nonlinear three-wave dynamics. It is also possible to infer that the absence of growth of the S waves should not to be regarded as indication of low efficiency of the decay process, since similar low level of excitation of S waves has also been obtained in the case of 1D analysis.[12]

Of particular significance in these reported results is the finding that the three-wave decay process in the present 2D situation apparently does not lead to the long wavelength Langmuir condensation, which was observed in the earlier 1D analysis of Ziebell *et al.*[12] The point is that the decay of primary Langmuir waves, L, can be said to be constituted by two processes. One involves the generation of back-scattered Langmuir waves, L', with similar wavelength but propagating in the anti-beam direction, while the other would involve the generation of small k, long wavelength, Langmuir condensate mode. The back-scattering process, $L \rightarrow L' + S$ is evidently quite effective, as Fig. 2 indicates. As a matter of fact, the back-scattering process seems to be even more robust in the present 2D case than in the 1D approximation. This can be attributed to the fact that, in 1D situation the L' mode generation is limited to the strictly anti-parallel mode, while, as Fig. 2 demonstrates, in the 2D case the L' mode can spread around the circular ring area in 2D wave number space. In contrast, the condensate mode (i.e., inverse cascade) generation is apparently less effective in the present 2D case. Further study must be made in order to assess if this behavior remains the same for different values of parameters ruling the instability, like the beam population, the beam velocity, and other plasma parameters.

It must be noted that, if the circular ring-like structure in 2D wave number space were to be projected onto 1D space along $k_\parallel$, the resulting reduced spectrum may lead to a false interpretation of Langmuir waves inverse cascading into long-wavelength regime. However, in reality, the actual wavelength does not increase at all, but only the wave propagation direction changes. This finding calls into question whether the well-established paradigm related to the late stages of the beam-generated Langmuir turbulence may have to be re-examined or not. The present results indicate that the strong turbulence scenario as described by Zakharov's equation, as final result of beam-generated Langmuir turbulence, may have to be re-examined.

However, we do not question the general validity of Zakharov theory in other contexts, such as when the Langmuir wave has very high and coherent amplitude, as in laser–plasma interaction or as in ionospheric heating problem.[46] It is simply that within the present context of the electron beam-generated Langmuir turbulence, the so-called inverse cascade of the condensation mode appears to be less effective than in the case of 1D approximation.

Of course, to further explore the 2D effects on Langmuir turbulence, one must include the other important nonlinear process that are known to lead to long-wavelength Langmuir condensation, namely, the spontaneous and induced scattering off thermal ions. One should also explore the parametric dependence of some features reported here. Another importance research tool would be to compare the present findings with full particle simulations in 2D. In this regard, the 2D simulation by Dum[47] is noteworthy. However, the physical parameters are quite different from those utilized in the present work, hence this calls for more detailed simulations in the future. These tasks obviously belong to our future work.

4. Conclusion

To recap and conclude the present chapter, we have solved the fully self-consistent weak turbulence equation, including nonlinear mode-coupling terms, in 2D for the first time. One of the findings is that the electron distribution undergoes perpendicular heating in addition to the formation of the quasilinear parallel plateau, as a result of electrostatic Langmuir/ion-sound turbulence process. This indicates a tendency to isotropization of the electron distribution function. This tendency to isotropization due to nonlinear processes may constitute an alternative mechanism for explanation to the observed quasi-isotropic energetic electrons in the solar wind, in contrast to the customary approach of invoking pitch-angle scattering by some sort of electromagnetic fluctuation, or collisional processes occurring near the sun.

Another significant finding is that the Langmuir turbulence in 2D situation forms a ring-like structure in wave number space, and apparently does not lead to condensation effects (at least under the present parameters, and for the present approximation of excluding scattering terms in the wave equation), in contrast to our earlier 1D analysis.[12] This calls for more detailed comparative analysis of Langmuir wave cascade process in 1D versus 2D situation. The standard paradigm in the community is that Langmuir condensation is a necessary precursor to Zakharov's strong Langmuir turbulence development. In view of this, further study of 2D Langmuir turbulence dynamics including the spontaneous and induced scattering (which the present chapter did not consider, as noted already) may lead to a much better understanding of not only the beam–plasma interaction process but also may shed light on strong turbulence effects.

Acknowledgments

This work has been partially supported by the Brazilian agencies Conselho Nacional de Desenvolvimento Científico e Tecnológico (CNPq) and Fundação de Amparo à Pesquisa do Estado do Rio Grande do Sul (FAPERGS). The work at the University of Maryland was supported by the National Science Foundation under grants ATM 0341084 and ATM0535821.

References

1. M. V. Goldman, *Sol. Phys.* **89** (1983) 403.
2. D. B. Melrose, in *Solar Radiophysics*, eds. D. J. McLean and N. R. Labrum (Cambridge Univ. Press, New York, 1985), p. 37.
3. P. A. Robinson and I. H. Cairns, *Sol. Phys.* **181** (1998) 363, 395, 429.
4. A. J. Willes, P. A. Robinson and D. B. Melrose, *Phys. Plasmas* **3** (1996) 149.
5. B. Li, A. J. Willes, P. A. Robinson and I. H. Cairns, *Phys. Plasmas* **12** (2005) 012103.
6. B. Li, A. J. Willes, P. A. Robinson and I. H. Cairns, *Phys. Plasmas* **12** (2005) 052324.
7. B. Li, P. A. Robinson and I. H. Cairns, *Phys. Plasmas* **13** (2006) 092902.
8. V. E. Zakharov, *Sov. Phys. JETP* **35** (1972) 908.
9. M. V. Goldman, *Rev. Modern Phys.* **56** (1984) 709.
10. P. A. Robinson, *Rev. Modern Phys.* **69** (1997) 507.
11. L. Muschietti and C. T. Dum, *Phys. Fluids B* **3** (1991) 1968.
12. L. F. Ziebell, R. Gaelzer and P. H. Yoon, *Phys. Plasmas* **8** (2001) 3982.
13. E. P. Kontar and H. L. Pécseli, *Phys. Rev. E* **65** (2002) 066408.
14. B. Li, A. J. Willes, P. A. Robinson and I. H. Cairns, *Phys. Plasmas* **10** (2003) 2748.
15. D. A. Russel, D. F. DuBois and H. A. Rose, *Phys. Rev. Lett.* **60** (1988) 581.
16. D. L. Newman, P. A. Robinson and M. V. Goldman, *Phys. Rev. Lett.* **62** (1989) 2132.
17. M. D. Montgomery, S. J. Bame and A. J. Hundhausen, *J. Geophys. Res.* **73** (1968) 4999.
18. W. C. Feldman, J. R. Asbridge, S. J. Bame, M. D. Montgomery and S. P. Gary, *J. Geophys. Res.* **80** (1975) 4181.
19. R. P. Lin, D. W. Potter, D. A. Gurnett and F. L. Scarf, *Astrophys. J.* **251** (1981) 364.
20. R. P. Lin, W. K. Levedahl, W. Lotko, D. A. Gurnett and F. L. Scarf, *Astrophys. J.* **308** (1986) 954.
21. W. G. Pilipp, H. Miggennieder, M. S. Montgomery, K.-H. Mühläuser, H. Rosenbauer and R. Schwenn, *J. Geophys. Res.* **92** (1987) 1075.
22. R. J. Fitzenreiter, J. D. Scudder and A. J. Klimas, *J. Geophys. Res.* **95** (1990) 4155.
23. R. J. Fitzenreiter, K. W. Ogilvie, D. J. Chornay and J. Keller, *Geophys. Res. Lett.* **25** (1998) 249.
24. R. E. Ergun *et al.*, *Astrophys. J.* **503** (1998) 435.
25. K. W. Ogilvie, L. F. Burlaga, D. J. Chronay and R. Fitzenreiter, *J. Geophys. Res.* **104** (1999) 22389.
26. C. Pagel, S. P. Gary, C. A. de Koning, R. M. Skoug and J. T. Steinberg, *J. Geophys. Res.* **112** (2007) A04103.
27. S. Saito and S. P. Gary, *J. Geophys. Res.* **112** (2007) A06116.
28. J. D. Scudder and S. Olbert, *J. Geophys. Res.* **84** (1979) 2755.
29. J. D. Scudder and S. Olbert, *J. Geophys. Res.* **84** (1979) 6603.
30. M. V. Canullo, A. Costa and C. F. Fontáin, *Astrophys. J.* **462** (1996) 1005.

31. Ø. Lie-Svendsen, V. H. Hansteen and E. Leer, *J. Geophys. Res.* **102** (1997) 4701.
32. V. Pierrard, M. Maksimovic and J. Lemaire, *J. Geophys. Res.* **104** (1999) 17021.
33. V. Pierrard, M. Maksimovic and J. Lemaire, *Astrophys. Space Sci.* **277** (2001) 195.
34. V. Pierrard, M. Maksimovic and J. Lemaire, *J. Geophys. Res.* **106** (2001) 29305.
35. S. Landi and F. G. E. Pantellini, *Astron. Astrophys.* **372** (2001) 686.
36. J. C. Dorelli and J. D. Scudder, *J. Geophys. Res.* **108**(A7) (2003) 1294, doi: 10.1029/2002JA009484.
37. C. Vocks and G. Mann, *Astrophys. J.* **593** (2003) 1134.
38. C. Vocks, C. Salem, R. P. Lin and G. Mann, *Astrophys. J.* **627** (2005) 540.
39. M. Maksimovic *et al.*, *J. Geophys. Res.* **110** (2005) A09104, doi: 10.1029/2005JA0111192005.
40. K. Appert, T. M. Tran and J. Vaclavik, *Phys. Rev. Lett.* **37** (1976) 502.
41. O. Ishihara and A. Hirose, *Phys. Rev. Lett.* **46** (1981) 771.
42. O. Ishihara and A. Hirose, *Phys. Fluids* **26** (1983) 100.
43. O. Ishihara and A. Hirose, *Phys. Fluids* **50** (1983) 1783.
44. L. Muschietti, I. Roth and G. Delory, *J. Geophys. Res.* **102** (1997) 27217.
45. P. H. Yoon, T. Rhee and C.-M. Ryu, *J. Geophys. Res.* **111** (2006) A09106.
46. J. G. Wang, G. L. Payne, D. F. Dubois and H. A. Rose, *Phys. Plasmas* **3** (1996) 111.
47. C. T. Dum, *Phys. Plasmas* **1** (1994) 1821.

Advances in Geosciences
Vol. 14: Solar Terrestrial (2007)
Eds. Marc Duldig *et al.*
© World Scientific Publishing Company

SUPERTHERMAL ELECTRON DISTRIBUTIONS
IN THE SOLAR WIND ENVIRONMENT

R. GAELZER

Instituto de Física e Matemática, UFPel, Caixa Postal 354
Campus UFPel, 96010-900 Pelotas, RS, Brazil
rudi@ufpel.edu.br

P. H. YOON

IPST, University of Maryland
College Park, Maryland 20742, USA

Pohang University of Science and Technology
Pohang, South Korea

L. F. ZIEBELL

Instituto de Física, UFRGS
Caixa Postal 15051, 91501-970 Porto Alegre, RS, Brazil

A. F. VIÑAS

NASA Goddard Space Flight Center
Greenbelt, Maryland 20771, USA

Electron distributions with various degrees of asymmetry associated with
the energetic tail population are commonly detected in the solar wind near
1 AU. By numerically solving one-dimensional electrostatic weak turbulence
equations, the present chapter demonstrates that a wide variety of asymmetric
energetic tail distribution may result. It is found that a wide variety of
asymmetric tail formation becomes possible if one posits that the solar wind
electrons are initially composed of thermal core plus field-aligned counter-
streaming beams, instead of the customary thermal population plus a single
beam. It is shown that the resulting nonlinear wave–wave and wave–particle
interactions lead to asymmetric non-thermal tails when there is a small
difference in the average beam speeds associated with the forward versus
backward components, whereas a large difference between the beam speeds
will generate a more symmetric tail in the electron distribution.

1. Introduction

The electron velocity distributions detected in the solar wind near 1 AU
are typically observed to be made of low energy and dense thermal core

plus two tenuous but hot superthermal populations, the *halo*, which is present at all pitch angles and the *strahl*, which is a highly anisotropic, field-aligned population.[1-12] In the literature, such particle distributions with thermal core plus the energetic tail are often modeled by the so-called kappa distribution[13-16] or by a combination of Maxwellian and bi-Maxwellian distributions when the strahl component is absent. However, the actual electron distributions measured in the solar wind (and to some degree, in the upstream region of the bow shock) often feature highly asymmetric forms such that they cannot be simply fit by model distributions.

The physical origin of the strahl or the isotropic halo electron populations which can be clearly distinguished from the Maxwellian core distribution remains as somewhat of a mystery to this day. In the literature, most theories that attempt to address the origin of superthermal electrons usually start from the consideration of the altitude-dependent collisional dynamics.[11,17-27] These classical approaches consider the Coulomb collisionality and the conservation of the first adiabatic invariant as the electrons propagate along open magnetic field with decreasing strength, as the main driving forces and effects behind the formation of the core and superthermal electrons. However, it has been long established that the strahl distribution is broader than it is expected from conservation of the adiabatic invariant alone, an effect that cannot be fully explained by Coulomb collisions, since that electron population at the spacecraft site is largely collisionless.[6,28]

In an attempt to explain the formation of both the halo and strahl populations, as well as the observed pitch angle diffusion of the strahl electrons, Vocks and Mann[25] and Vocks *et al.*[26] studied the quasilinear diffusion of a kappa distribution function in the presence of whistler waves propagating both in the sunward and anti-sunward directions. Although they were able to obtain distributions that resemble the observed ones, their model was not fully self-consistent. In particular, they did not consider the Landau damping on the whistler waves, which should limit the total energy of the resonant wave available for pitch-angle scattering.[29] Moreover, since the observed distributions are measured with time intervals that can range from several seconds up to minutes, depending on the pitch angle resolution, higher-order nonlinear effects, such as wave–wave and nonlinear wave–particle interactions, should be included for a complete kinetic description.

In contrast to these theories, in the present chapters we look for collective mechanisms which include nonlinear effects as a potential explanation for the generation of superthermal population. As the solar wind expands into the surrounding interplanetary medium, the faster electrons outpacing the slower ones will inevitably lead to the formation of the field-aligned beams. The beam electrons will excite plasma instability which will in turn slow down the beams. In short, there will be a constant "struggle" between the quasilinear relaxation and the time-of-flight beam reformation process. Thus, we expect that, while collisional dynamics may be important very close to the solar surface, in most of the interplanetary space, the collective dynamical processes may play an important role on the electron scattering processes. We shall thus investigate the role of plasma instability and turbulent dynamical processes on the formation of the superthermal tail population.

The basic demonstration that turbulent processes leads to these hot plasma populations has already been accomplished in the papers by Yoon *et al.*[30] and Rhee *et al.*[31] The focus of the present discussion is on the question of what leads to the observed asymmetry associated with the solar wind superthermal electrons. In the present chapter, we thus put forth a theory, in which the highly asymmetric energetic tail distribution may be the result of dense thermal core electrons interacting with not just one component of field-aligned energetic electron beam component, but rather a pair of initially counter-streaming energetic but tenuous electron beam populations. The resulting nonlinear interactions among the electrons with the Langmuir and ion-sound turbulence leads to the observed highly asymmetric superthermal tail population.

To demonstrate this, we have carried out a series of numerical analyses in which self-consistent weak turbulence equations are solved for initial electron distribution consisting of three components — the background plus the forward and backward traveling electron beam populations. The kinetic equations for waves include the effects of spontaneous and induced emissions, three-wave decay and nonlinear scattering terms, whereas the kinetic equation for the electron distribution function contains the contributions of spontaneous drag and quasilinear terms. It is shown that the resulting development of Langmuir turbulence excited by the interaction among the counter-streaming electrons and the background thermal distribution of electrons leads to the asymmetric energetic tail distributions. From this, we suggest that the observed asymmetric superthermal electrons in the solar wind may be the result of turbulence acceleration by self-consistently generated Langmuir waves excited by the three-component electron distribution system associated with the solar wind.

The plan of the chapter is as follows. In Sec. 2 some observational results from the WIND spacecraft are presented and discussed. In the context of these observations, Sec. 3 presents the basic theory of weak plasma turbulence and the model adopted in this work, followed by the main results obtained from the model. Finally, Sec. 4 discusses the results and shows our conclusions.

2. WIND/SWE-VEIS Electron Observations

The electron 3D-velocity distribution function (VDF) measurements used in our study have been obtained from the 3-s time resolution data of the WIND/SWE VEIS-spectrometer (details of the instrument characteristics can be found in Ref. 32. The VEIS spectrometer consists of six programmable analyzers which forms a three-pair of mutually orthogonal sensors that measures electrons in the energy range from 7 eV to 25 keV in 16 energy steps with an energy resolution of about 6%. However, for solar wind electron studies the effective energy range has been set from 10 eV to 3 keV. Each sensor full energy sweep takes 0.5 s, which implies that the highest time resolution moment is determined in 0.5 s, but for statistical purposes, the moments have been averaged out to the full satellite spin period of 3 s. The moment calculations has been corrected by the spacecraft potential (which usually range between 3 and 15 V depending upon the solar conditions) using either the proton and alpha measurements from the WIND/SWE MIT Faraday Cup or the electron density estimates of the plasma frequency line from the WIND/WAVES experiment.

The VDF presented in this chapter have been shifted into the solar wind frame using the proton bulk velocity interpolated to the electron times and the higher order moments (e.g., $\mathbf{P}_e$ and $\mathbf{Q}_e$) estimated have been calculated in that frame. The VDF data sets used in this study are the reduced $F(v_\parallel)$ distribution functions obtained by folding the original 3D distributions into the $(v_\parallel, v_\perp)$ space (e.g., assuming the gyrotropy condition) using the measured 3-s magnetic field averages and then integrating the 2D distributions in the $v_\perp$ space.

Figures 1 and 2 show two samples of typical reduced electron velocity distribution functions during a long-orbit excursion of the WIND spacecraft to L1 in the solar wind, which occurred on 1 February, 1995. The plots also display fits (solid line) to the observed distributions (asterisk) using a model of a superposition of a Maxwellian (core) and a kappa-distribution (halo) models. The electron VDF shown in Fig. 1 clearly shows asymmetric pronounced tails along the magnetic field, which is an indication of the

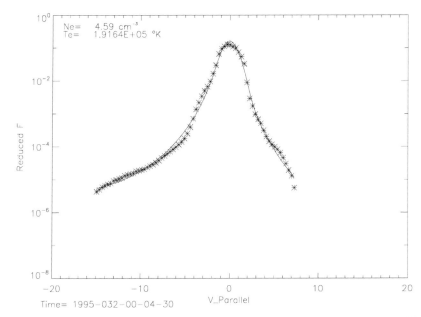

Fig. 1. A sample of electron distribution detected in the solar wind featuring highly asymmetric tail population.

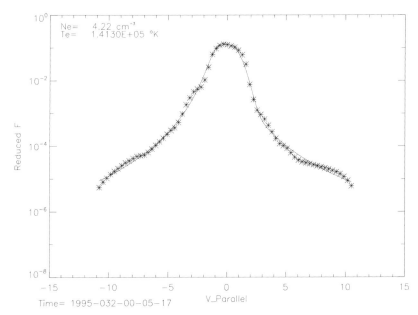

Fig. 2. A sample electron distribution detected in the solar wind featuring quasi-symmetric tail population.

heat-flux. The parallel velocity $(v_{\parallel})$ is obtained by decomposing the full velocity vector $\mathbf{v}$ relative to the local magnetic field $\mathbf{B}$, so that $\mathbf{v}_{\parallel} = \mathbf{v} \cdot \mathbf{B}/|\mathbf{B}|$ in the GSE coordinate system, whose x-component points from the Earth to the Sun. In contrast, the electron VDF displayed in Fig. 2 features quasi-symmetric superthermal tail population. For the two VDF's selected in Figs. 1 and 2, the B-field components are: $\mathbf{B} = (4.560, -2.355, 2.352)$ nT, and $\mathbf{B} = (4.812, -2.910, -1.014)$ nT, respectively. Notice that the fields are mostly in the x-direction, which in GSE is pointing toward the Sun, but the heat-flux is mainly from the Sun to the Earth. Therefore, we expect the distributions to be skewed in the negative-$v_{\parallel}$ direction as shown in Figs. 1 and 2.

It is not possible to distinguish between halo and strahl electrons in the reduced 1D distributions presented in Figs. 1 and 2, as it would be in the 2D distributions. As a result of the reduction process, the superthermal tails presented are formed with the combination of both populations. However, the reduced distribution is better suited to our purposes, since our theory is still restricted to 1D and it does not consider the different physical origins of halo and strahl electrons, but focuses on the nonlinear evolution of the superthermal tail fueled by the free energy provided by electron beams proposed for the initial-time distribution.

3. Theory and Results

The basic theoretical framework is the self-consistent weak turbulence equations that describe nonlinear interactions among Langmuir and ion-sound modes as well as with the particles. In the present analysis, we resort to 1D approximation and assume that only electrostatic interaction is of importance. The equations of weak turbulence theory can be found in the papers by Yoon et al.[30] and Rhee et al.[31] These consist of Langmuir and ion-sound wave kinetic equations and the particle (electron) kinetic equation. Since the basic equations to be numerically solved are the same as in the paper by Yoon et al.,[30] we shall not repeat them here. We simply note that the normalization of the physical quantities in the present chapter is such that dimensionless time, speed, and the wave number are given by

$$T = \omega_{pe}\, t,$$

$$v \to v/v_{Te}, \tag{1}$$

$$k \to k v_{Te}/\omega_{pe}.$$

That is, by simply referring to v and k, we mean, respectively, the normalized quantities as defined above. In (1), $\omega_{pe} = \sqrt{4\pi\hat{n}e^2/m_e}$ is the plasma frequency, $\hat{n}$, e, and m_e being the ambient density, unit electric charge, and electron mass, respectively; v is a 1D velocity; $v_{Te} = \sqrt{2T_e/m_e}$ is the electron thermal speed, T_e being the electron temperature; and k is the 1D wave number.

The basic physics of self-consistent theory of superthermal tail generation during the beam–plasma interaction has been elucidated in the papers by Yoon *et al.*[30,33] and Rhee *et al.*[31] In these papers, the interaction of an initial Gaussian distribution of energetic electron beam with thermal electrons is considered. It was shown that the final asymptotic electron distribution resembled the well-known kappa distribution[13–16] with a non-Gaussian energetic tail. The physical mechanism responsible for the acceleration of electrons was also identified as the so-called spontaneous scattering of electrons off thermal ions, mediated by Langmuir turbulence of intermediate wavelengths. The spontaneous scattering process is a nonlinear wave–particle interaction that is operative only when the discrete-particle nature of the plasma is taken into account.

In spite of the fact that basic underlying physics of superthermal electron formation is understood, the theory discussed in the papers by Yoon *et al.*[30,33] and Rhee *et al.*[31] has limitation in that the asymptotic quasi-kappa distribution that emerges from such a theory is more or less invariant in terms of the degree of asymmetry associated with the forward-going (positive velocity) versus backward-propagating (negative velocity) particles. In order to explain a wide variety of asymmetric energetic tail distribution typically observed in the solar wind, one must modify the fundamental assumptions.

In view of the fact that modeling the solar wind as a core Maxwellian plus a single beam does not give us enough freedom to generate a wide variety of asymmetric tail distribution, we have decided to allow for the presence of a small but oppositely directed secondary beam component. We have thus revisited our previous model considered by Yoon *et al.*[30,33] and Rhee *et al.*,[31] and replaced the initial electron distribution with a different initial configuration in which a secondary backward-propagating Gaussian beam is added to the original configuration of Gaussian beam plus the background.

The justification for the presence of counter-streaming electrons is more than academic. At 1 AU, the interplanetary magnetic field structure may be rather complicated. In some special circumstances, when both foot

points of the field line loops are located on the Sun, the field-aligned
motion of the electrons may indeed be characterized as counter-streaming.
Sunward-propagating beams can also be observed inside closed magnetic
field structures, which can be generated by flare-induced interplanetary
shock waves.[6] Also, on some rare occasions, when the interplanetary
field lines are tangent to the planetary bow shock and the spacecraft is
sufficiently close, the reflected electrons off the cross-shock electrostatic
potential may coexist with the incoming solar wind electrons. In such
situations, the assumption of counter-streaming population is again
justifiable.

As we shall demonstrate subsequently, the resulting numerical solutions
with the initial counter-streaming electrons plus the dense core population
are quite extraordinary in that a wide-ranging asymmetry associated with
the energetic electron tail distribution can be obtained, which could not
have been foreseen *a priori*. In what follows, we showcase two extreme
situations. One case will correspond to a highly asymmetric energetic tail
distribution, while in the second case we shall demonstrate symmetric
tail distribution. However, we hasten to point out that we were also able
to generate a wide range of asymmetric distributions between the two
extremes.

The input physical parameters are the ratio of the forward-propagating
electron beam density to the background number density, n_f; ratio of the
backward-propagating electron beam density to the background number
density, n_b; ratio of forward and backward beam speed to thermal speed,
U_f and U_b; ion-to-electron temperature ratio, τ; ion-to-electron mass ratio,
M; and the plasma parameter g:

$$n_f = \hat{n}_f/\hat{n}_0,$$
$$n_b = \hat{n}_b/\hat{n}_0,$$
$$U_f = V_f/v_{Te},$$
$$U_b = V_b/v_{Te}, \tag{2}$$
$$\tau = T_i/T_e,$$
$$M = m_i/m_e,$$
$$g = 1/\hat{n}\,\lambda_D^3.$$

Here $\lambda_D^2 = T_e/(4\pi\hat{n}e^2)$ is the square of the electron Debye length. In the
present analysis, we assume that Gaussian thermal spreads associated with
the core electrons, and the counter-streaming beams are the same. Of the

above, we assume that $\tau = 1/7$, $m = 1836$, and $g = 5 \times 10^{-3}$. We do not vary these quantities in the subsequent numerical analysis.

The initial electron distribution and stationary ion distribution are given respectively, in normalized form, by

$$
\begin{aligned}
f_e(v,0) &= \pi^{-1/2} \left\{ \exp\left(-v^2\right) + n_f \exp\left[-(v - U_f)^2\right] \right. \\
&\quad \left. + n_b \exp\left[-(v - U_b)^2\right] \right\},
\end{aligned}
\tag{3}
$$

$$
f_i(v) = \sqrt{\frac{M}{\pi\tau}} \exp\left(-\frac{Mv^2}{\tau}\right).
$$

The turbulence intensity is normalized according to the convention

$$
I_{\sigma\alpha}(k) = \frac{g\, I_k^{\sigma\alpha}}{8\sqrt{2}\, m_e\, v_{Te}^2},
\tag{4}
$$

where $I_k^{\pm\alpha}/(8\pi) = \left\langle \delta E_{\pm}^{\alpha 2} \right\rangle_k /(8\pi)$ is the ensemble average of the spectral wave electric field energy density corresponding to mode α propagating in the forward/backward direction with respect to the beam propagation direction. The initial perturbation is specified according to the spontaneous emission formula, and is given by

$$
\begin{aligned}
I_{\pm L}(k,0) &= \frac{g}{2^6\sqrt{2}\,\pi^2} \frac{1}{1 + 3k^2/2}, \\
I_{\pm S}(k,0) &= \frac{g\,k^2}{2^7\sqrt{2}\,\pi^2} \sqrt{\frac{1 + k^2/2}{1 + 3k^2/2}} \\
&\quad \times \left. \frac{e^{-v^2} + (M/\tau)^{1/2} e^{-Mv^2/\tau}}{e^{-v^2} + \tau^{-1}(M/\tau)^{1/2} e^{-Mv^2/\tau}} \right|_{v = v_r}, \\
v_r &= \sqrt{\frac{1 + 3\tau}{M(2 + k^2)}}.
\end{aligned}
\tag{5}
$$

We have solved the weak turbulence equation with the standard leap-frog method with the time increment $\Delta T = 0.01$. We take 201 velocity grid points and 101 grid points for the wave numbers. The wave intensities are computed over the positive k space, but we plot backward-propagating wave intensities, $I_{-L}(k)$ and $I_{-S}(k)$, over the negative k space by invoking the symmetry relations, $I_{\pm L}(k) = I_{\mp L}(-k)$ and $I_{\pm S}(k) = I_{\mp S}(-k)$.

3.1. Asymmetric tail

In the first example, we consider the following set of input parameters:

$$U_f = 4,$$
$$U_b = -3,$$
$$n_f = 1 \times 10^{-2}, \tag{6}$$
$$n_b = 1 \times 10^{-3}.$$

Figure 3 shows the time evolution of the total electron distribution $f_e(v)$ versus v (top), and Langmuir (bottom-left) and ion-sound (bottom-right) mode spectral wave energy intensities, $I_L(k)$ and $I_S(k)$, versus k. The final time step corresponds to normalized time $T = 1 \times 10^4$, with every $\Delta T = 1,000$ intermediate time step plotted with dashes. For the present

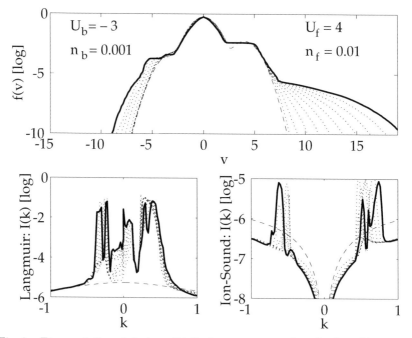

Fig. 3. Time evolution of electron distribution versus $v = v/v_{Te}$ (top), and Langmuir (bottom-left) and ion-sound (bottom-right) mode spectral wave energy intensities versus $k = kv_{Te}/\omega_{pe}$. The final time step corresponds to normalized time $T = 1 \times 10^4$, with every $\Delta T = 1,000$ intermediate time step plotted with dashes. Forward beam is characterized by $n_f = 0.01$ and $U_f = 4$, while the parameters for the backward beam are $n_b = 0.001$ and $U_b = -3$.

choice of the backward beam speed $U_b = -3$, which is slightly less than the forward beam speed, $U_f = 4$, and for the forward beam density sufficiently higher than the backward beam density, $n_f = 0.01$ and $n_b = 0.001$, we find that the asymptotic state of the electron distribution function is a highly asymmetric in that the positive v range features an extended higher-energy tail, but the negative v space has virtually no energetic tail distribution.

To see whether the highly-asymmetric tail formation is the result of the differences in the magnitude of beam speed or the density, we have varied U_b and n_b separately, but we found that it is the beam speed that determines the degree of asymmetry associated with the energetic tail. To show this, we display in Fig. 4, the results with the same input parameters as in Fig. 3 — see Eq. (6), except that

$$n_b = 0.005. \tag{7}$$

Note that the overall feature associated with the electron distribution function is qualitatively similar to Fig. 2 in that positive v space has

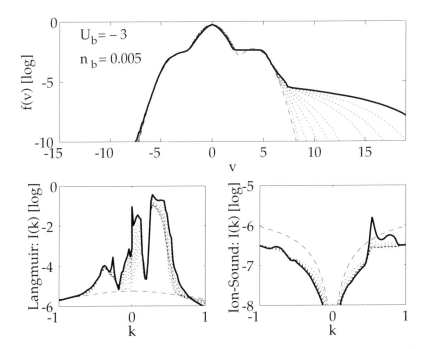

Fig. 4.   The same as Fig. 3, except that the backward beam density is raised to 0.5%.

extended tail while the negative v has no energetic non-Gaussian tail. As a matter of fact, the slight increase in the backward beam density has resulted in the suppression of what little non-Gaussian feature the electrons with negative v had before. The increase in n_b has concomitantly resulted in a noticeable increase in the population of energetic electrons with positive v.

To further test the hypothesis that the change in the backward beam density does not qualitatively alter the asymmetry associated with the energetic tail distribution, we now consider the equal density counter-streaming beam situation, i.e.,

$$n_b = 0.01. \tag{8}$$

The numerical solutions are shown in Fig. 5. As the reader may appreciate, the results are virtually the same as Fig. 4. From this, we tentatively conclude that the asymmetry in the counter-streaming beam speeds is responsible for generating asymmetric energetic tails.

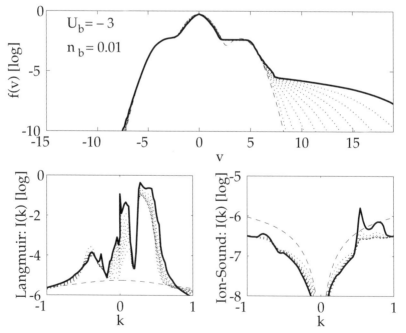

Fig. 5.   The same as Fig. 3, except that the backward beam density is further raised to 1%.

3.2. *Symmetric tail*

To further confirm the above tentative finding, we now consider the equal beam speed counter-streaming situation with the beam parameters,

$$\begin{aligned} U_f &= 4, \\ U_b &= -4, \\ n_f &= 1 \times 10^{-2}, \\ n_b &= 1 \times 10^{-3}. \end{aligned} \tag{9}$$

Figure 6 is in the same format as Fig. 3. For the present choice of the beam speeds $U_b = -4$ and $U_f = 4$, and for $n_f = 0.01$ and $n_b = 0.001$, we find that the asymptotic state of the electron distribution function features a quasi-symmetric high-energy tail population.

We now increase the backward beam density to

$$n_b = 0.005. \tag{10}$$

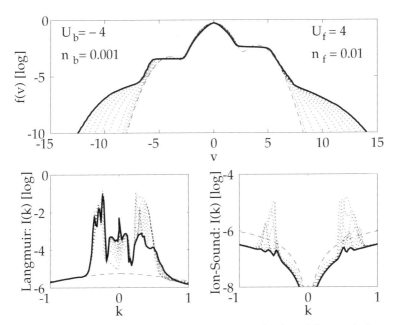

Fig. 6. Time evolution of electron distribution versus v (top), and Langmuir (bottom-left) and ion-sound (bottom-right) mode spectral wave energy intensities versus k. The initial counter-streaming beam parameters are $n_f = 0.01$ and $U_f = 4$, while the parameters for the backward beam are $n_b = 0.001$ and $U_b = -3$.

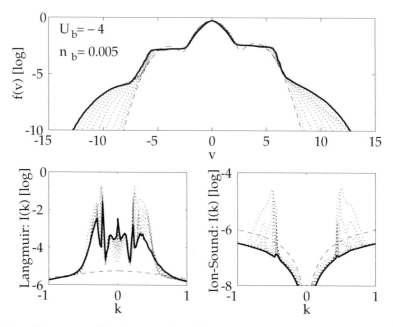

Fig. 7.   The same as Fig. 6, except that the backward beam density is raised to 0.5%.

The numerical result is displayed in Fig. 7. As one can see, the degree of symmetry in the forward ($v > 0$) and backward ($v < 0$) propagating electron tail populations has somewhat increased when compared with Fig. 6.

Finally, we move on to the case of equal counter-streaming beam speeds and densities, namely,

$$n_b = 0.01. \tag{11}$$

The numerical solutions shown in Fig. 8 correspond to highly symmetric energetic tail distribution for the particles and symmetric wave spectra.

Although we have chosen only two representative cases, we have actually considered other cases as well. Depending on the backward beam speed U_b, we found that the degree of asymmetry associated with the energetic tail greatly varies. However, the dependence of the asymmetry on U_b is not a simple linear relationship. For instance, as the previous Figs. 3–8 show, the case of $U_b = -4$ led to symmetric two-sided tail distribution, whereas the case of $U_b = -3$ produced an almost one-sided tail. From this, one may naively expect that a further reduction of U_b from $U_b = -3$ to

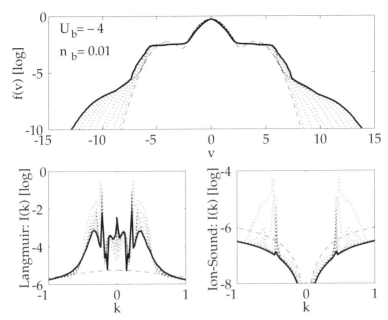

Fig. 8. The same as Fig. 6, except that the backward beam density is further raised to 1%. This situation corresponds to equal counter-streaming beam speeds and densities.

say, $U_b = -2.5$, may lead to even more prominent one-sided tail in the positive v direction. However, this is not the case at all. In fact, we found that the reduction of $U_b = -3$ to something like $U_b = -2.5$ actually caused the negative v tail to grow back. The case of $U_b = -3$ actually corresponds to the extremum case where the energetic tail is almost completely one-sided. This finding suggests that the energization of the electrons and the formation of asymmetric tail population is not a simple phenomenon to be explained intuitively.

The main reason for the difference between the asymmetric and the symmetric tails can be observed in the bottom left panels of Figs. 3–8, which shows the time evolution of the Langmuir waves. In the case of symmetric tail, the spectrum is completely symmetrical for forward- and backward-propagating Langmuir waves, particularly in the small wave number region, where the wave intensity grows from $T = 0$ basically due to nonlinear scattering with the particles situated on the tail of the distribution function. Observing now the asymmetric case in Figs. 3–5, one immediately notices that the Langmuir spectrum is asymmetric in the small wave number region as well, that is, forward-propagating waves are more efficiently enhanced in this region due to the nonlinear scattering process than backward-propagating waves. As a result, particles are more efficiently scattered to a single side of the distribution, generating the asymmetric tail. This asymmetry is observed only when there is a slight mismatch between the forward and backward beam speeds, as in the $U_f = 4$ and $U_b = -3$ case. As the difference in the absolute values of the speeds increase, the nonlinear scattering process becomes ever more symmetric again. This is an unexpected result from the nonlinear physics of weak turbulence processes, which we believe can be responsible for at least a part of asymmetric distributions observed in the solar wind.

Finally, an important order-of-magnitude analysis can be made concerning the typical amount of time required for a substantial high-energy tail to form. The final VDFs displayed by our results show that a quasi-steady state forms for a normalized time of $T = 10^4$. Assuming an electron density of $n_e \simeq 5 \, \text{cm}^3$, usually observed by the Wind spacecraft (see Figs. 1 and 2), we have a typical plasma frequency of 20 kHz in the solar wind at 1 AU. This means that the final distribution functions we obtained correspond to a total real time around 0.08 s, smaller by a factor of at least 10 than the spin period of 3 s of the Wind spacecraft (see discussion in Sec. 2). Therefore, it is reasonable to assume that the observed VDF's contain a large amount of nonlinearly scattered particles, as one would expect from our theory.

4. Conclusions and Discussion

The electron velocity distribution functions (VDFs) detected in the solar wind near 1 AU are typically observed to be made of a thermal core plus superthermal populations. The different electron populations at a given point in interplanetary space usually come from distinct sources, such as the solar chromosphere or the bow shock, for instance. The physical origin and the evolution of the strahl and halo populations are still not entirely understood to this day. In the literature, most theories rely on altitude-dependent collisional dynamics. However in our recent works,[30,31] we put forth an alternative mechanism for the generation of superthermal particles. In this view, collective turbulent processes are responsible for the acceleration of electrons to superthermal energies. By solving weak turbulence kinetic equations, we have demonstrated that the self-consistent electron VDFs indeed feature kappa-like superthermal energetic tails, thus confirming that the turbulent acceleration is a viable explanation for the generation of superthermal core population.

The motivation for the present discussion stems from the fact that the actual electron VDFs measured in the solar wind (and to some degree, in the upstream region of the bow shock) often feature highly asymmetric forms such that they cannot be simply fit by bi-Maxwellians or kappa models. Despite their different physical origins, the electron populations present at a given point in the solar wind should interact through linear and nonlinear mechanisms and the resulting VDF measured by the spacecraft at the end of its sampling period, which is typically of some seconds up to a few minutes, should contain a fair amount of nonlinearly scattered particles. Thus, in the present chapter we have generalized our theory to include a pair of initially counter-streaming energetic but tenuous electron beam populations. On the basis of such an initial configuration, we have demonstrated that a wide variety of asymmetric energetic tail distribution may result. It is found that the asymmetric tail distribution is reminiscent of typical electron distribution detected near 1 AU in the solar wind.

When the forward- and backward-propagating components of the tenuous energetic electrons interact with low energy core solar wind electrons, Langmuir turbulence is excited. The ensuing nonlinear wave–wave and wave–particle interactions are shown to lead to asymmetric high-energy tail distributions. By numerically solving nonlinear weak turbulence equations for electrons and Langmuir turbulence over a range of physical input parameters, we have identified that the delicate difference in the counter-streaming beam speeds is the main cause for generating the asymmetry in the energetic tail, and that the difference in the counter-streaming beam density ratio is immaterial. The present finding may provide a useful diagnostics for the *in situ* measurement of the solar wind electrons.

Our theory was restricted, in this work, to 1D electron distribution functions and wave spectra. Future works will generalize our results to 2D situations and our new results will be presented in the literature.

Acknowledgments

The research at the University of Maryland was supported by NSF grant ATM0535821. This work has been partially supported by the Brazilian agencies CNPq and FAPERGS.

References

1. M. D. Montgomery, S. J. Bame and A. J. Hundhaus, *J. Geophys. Res.* **73** (1968) 4999.
2. W. C. Feldman *et al.*, *J. Geophys. Res.* **80** (1975) 4181.
3. R. P. Lin *et al.*, *Astrophys. J.* **251** (1981) 364.
4. R. P. Lin *et al.*, *Astrophys. J.* **308** (1986) 954.
5. W. G. Pilipp *et al.*, *J. Geophys. Res.* **92** (1987) 1075.
6. W. G. Pilipp *et al.*, *J. Geophys. Res.* **92** (1987) 1093.
7. R. J. Fitzenreiter, J. D. Scudder and A. J. Klimas, *J. Geophys. Res.* **95** (1990) 4155.
8. R. J. Fitzenreiter *et al.*, *Geophys. Res. Lett.* **25** (1998) 249.
9. R. E. Ergun *et al.*, *Astrophys. J.* **503** (1998) 435.
10. K. W. Ogilvie *et al.*, *J. Geophys. Res.* **104** (1999) 22389.
11. V. Pierrard, M. Maksimovic and J. Lemaire, *Astrophys. Space Sci.* **277** (2001) 195.
12. C. Pagel *et al.*, *J. Geophys. Res.* **110** (2005) A01103.
13. V. M. Vasyliunas, *J. Geophys. Res.* **73** (1968) 2839.
14. D. Summers and R. M. Thorne, *Phys. Fluids B* **3** (1991) 1835.
15. R. L. Mace and M. A. Hellberg, *Phys. Plasmas* **2** (1995) 2098.
16. L. Yin *et al.*, *J. Geophys. Res.* **103** (1998) 29595.
17. J. D. Scudder and S. Olbert, *J. Geophys. Res.* **84** (1979) 6603.
18. J. D. Scudder and S. Olbert, *J. Geophys. Res.* **84** (1979) 2755.
19. M. V. Canullo, A. Costa and C. FerroFontan, *Astrophys. J.* **462** (1996) 1005.
20. Ø. Lie-Svendsen, V. H. Hansteen and E. Leer, *J. Geophys. Res.* **102** (1997) 4701.
21. V. Pierrard, M. Maksimovic and J. Lemaire, *J. Geophys. Res.* **104** (1999) 17021.
22. V. Pierrard, M. Maksimovic and J. Lemaire, *J. Geophys. Res.* **106** (2001) 305.
23. S. Landi and F. G. E. Pantellini, *Astron. Astrophys.* **372** (2001) 686.
24. J. C. Dorelli and J. D. Scudder, *J. Geophys. Res.* **108** (2003) 1294.
25. C. Vocks and G. Mann, *Astrophys. J.* **593** (2003) 1134.
26. C. Vocks *et al.*, *Astrophys. J.* **627** (2005) 540.
27. M. Maksimovic *et al.*, *J. Geophys. Res.* **110** (2005) A09104.
28. C. Pagel *et al.*, *J. Geophys. Res.* **112** (2007) A04103.
29. S. Saito and S. P. Gary, *J. Geophys. Res.* **112** (2007) A06116.
30. P. Yoon, T. Rhee and C. Ryu, *J. Geophys. Res.* **111** (2006) A09106.
31. T. Rhee, C. Ryu and P. Yoon, *J. Geophys. Res.* **111** (2006) A09107.
32. K. W. Ogilvie *et al.*, SWE, a comprehensive plasma instrument for the wind spacecraft, in *The Global Geospace Mission*, ed. C. T. Russel (Kluwer, Dordrecht, 1995), pp. 55–77.
33. P. H. Yoon, T. Rhee and C.-M. Ryu, *Phys. Rev. Lett.* **95** (2005) 215003.

Advances in Geosciences
Vol. 14: Solar Terrestrial (2007)
Eds. Marc Duldig *et al.*
© World Scientific Publishing Company

SOLAR RADIO BURST FINE STRUCTURES IN 1–7.6 GHz RANGES

YIHUA YAN

National Astronomical Observatories
Chinese Academy of Sciences, Beijing 100012, China

The State Key Laboratory of Space Weather
P. O. Box 8701, Beijing 100080, China
yyh@bao.ac.cn

The observations of the solar radio bursts with fine structures in decimetric–centimetric wave ranges provide important information on the fundamental physical processes of the primary energy release, particle acceleration and transportations in solar events. Using spectropolarimeters at 1–2 GHz, 2.6–3.8 GHz, and 5.2–7.6 GHz at NAOC/Huairou with very high temporal (1.25–8 ms) and spectral (4–20 MHz) resolutions, together with other observations, we are able to analyze a number of radio-fine structures including zebra patterns, microwave-type U-bursts, spikes, etc., so as to understand the nature of these processes and to diagnose coronal plasma parameters.

1. Introduction

Radio observations at decimetric- and centimeter-wavelengths are important for addressing fundamental processes of energy release, particle acceleration and particle transport in solar events.[1-3] Temporal fine structure (FS) of solar radio emission in solar flares has been found at various wavelengths for several decades. FSs such as spikes, pulsations, zebras, etc., provide important information for diagnosing coronal plasma parameters and the dynamic process in the solar corona.

Metric and kilometer type-III bursts satisfy an empirical relation for the decay time with frequency[4]:

$$\tau = 10^{7.71}\nu^{-0.95} \ (s). \tag{1}$$

From Eq. (1) one may estimate the decay times at different frequencies[5] as listed in Table 1.

Table 1. Estimation of decay times at different frequencies.

Frequency	0.3 GHz	1.42 GHz	2.84 GHz	5.7 GHz	8.0 GHz
Decay Time	454 ms	104 ms	54 ms	28 ms	20 ms

Table 2. Performance of the solar radio spectrograph at Huairou/NAOC.

ν (GHz)	$\Delta\nu$ (MHz)	Δt (ms)	S_{min} and Dyn. Range	Polari.	Operation
1–2	20	100	(2% ~ 10 dB) quiet $S_\odot$	R,L	1994–2003
	4	5	(2% ~ 10 dB) quiet $S_\odot$	R,L	2003–(upgrade)
2.6–3.8	10	8	(2% ~ 10 dB) quiet $S_\odot$	R,L	1996–
5.2–7.6	20	5	(2% ~ 10 dB) quiet $S_\odot$	R,L	1999–

The duration of spikes is comparable to the electron–ion collision time as noted in Tarnstrom and Philip.[6] The duration of individual FS decreases as the frequency increases. Consequently, at microwave frequencies, it is essential to have high temporal (Δt) and spectral ($\Delta\nu$) resolutions and high sensitivity (S_{min}) for observing FSs. The Chinese Solar Radio Broadband Fast Dynamic Spectrograph was developed to meet these requirements.[5] Table 2 shows the main parameters of the solar radio spectrometers at Huairou Station of NAOC.

2. Solar Radio Bursts

Solar Radio FSs in decimetric and higher frequency range (> 1GHz) have been cataloged into different types.[5,7,8] The occurrence of FSs with respect to the earth-effective CMEs was presented in Yan *et al.*[9]

With the spectrometers at Huairou/NAOC, many FSs have been observed, and Table 3 summarizes the bursts registered until 2006. In the following, we discuss some typical FSs.

2.1. Microwave type III bursts

Type III bursts are considered as plasma emission due to energetic electron beams along field lines.[10–13]

At microwaves, the frequency drift rates are mostly positive (from lower to higher frequencies) with 40 MHz/s–22 GHz/s. The duration at one frequency channel is from $<$30 ms to 200 ms. The generation mechanism

Table 3.   Solar radio bursts observed at Huairou/NAOC.

Frequency range	1–2 GHz (1994–2006)	2.6–3.8 GHz (1996–2006)	5.2–7.6 GHz (1999–2006)
Burst events	1131	2154	1428
Fine structures	210	166	54

of type IIIs at high frequency is not clear.[14] For example, how does radio emission effectively escape from high density plasma?

The occurrence of bi-directional type IIIs is considered to correspond to the primary acceleration site. The accelerated particles propagating along magnetic field lines both upward and downward may generate these bi-directional type III bursts. For the 5 January 1994 burst at 05:52:12UT, the parameters at the reconnection site were estimated with a height of 30 Mm,[15] an electric field intensity 10^{-4} V/m, length 10 Mm,[16] plasma $\beta \sim 0.01$, and ambient magnetic field $\sim$100 G.[17] On 13 December 2006, two bi-directional type IIIs were observed, as shown in Fig. 1.

2.2. *Microwave U-, N-, and M-shaped bursts*

Electron beams moving along a magnetic loop with density minimum at the loop top may cause U-shaped bursts at meter–decimetric wavelengths due to the beam instability and kinematics. If the beam is reflected at the magnetic mirror point, N-shaped or M-shaped bursts may be observed. Plasma parameters are stationary in this case. In the microwave range, similar structures have been recorded as shown in Figs. 2 and 3.

For two U-shaped bursts shown in right panels of Fig. 2, we could locate the radio sources at both footpoints with the SSRT observations. The source size along the loop is only of the order of a few Mm and does not show a large separation ($>$30 Mm) between sources at different branches if it would be a loop. In cm-wavelengths U-structures should be produced by density variations due to the plasma response to a heating pulse either by loss-cone instability or MHD time-dependent process.[18] There are two aspects to explain these results. One is due to the bounce period of the short electron beam in the long magnetic loop. From the lifetime duration the beam velocity must be 0.45 c and the loop length about 20 Mm. The other is due to the transverse MHD oscillations of the loop (for $B = 100$ G, diameter of the loop must be about 100 km).[18]

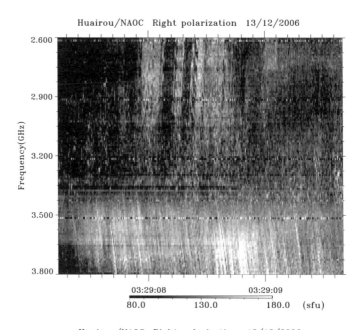

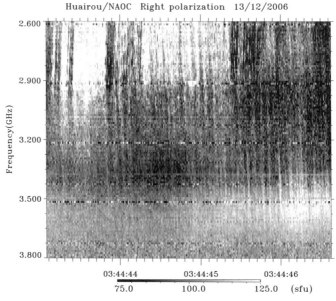

Fig. 1.   The bi-directional Type-III bursts on 13 December 2006 (a) at 03:29 UT
with frequency drift rates of about $-10\,\mathrm{GHz/s}$ (upward) and $+5\,\mathrm{GHz/s}$ (downward);
and (b) at 03:44 UT with drift rates of about $-13\,\mathrm{GHz/s}$ (upward) and $+10\,\mathrm{GHz/s}$
(downward).

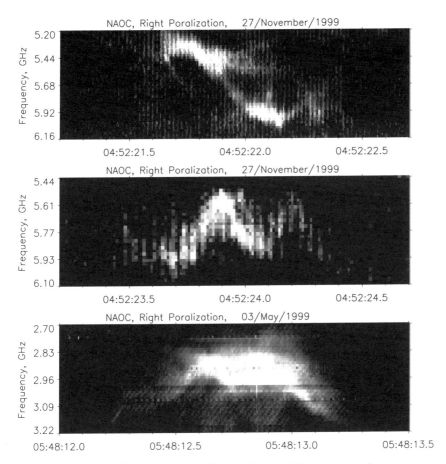

Fig. 2. The microwave J-, N-, and M-shaped bursts observed.

2.3. *Microwave zebras and fibers, with super-FSs*

The more or less regular stripes of emission and absorption, known as zebra-pattern or fiber bursts with an intermediate frequency drift against the continuum emission of type-IV radio bursts at meter or decimeter wavelengths are well known in the literature.[19–23] With Huairou/NAOC 2.6–3.8 GHz spectrograph, it demonstrates that this type of fine structure displays the same variety at centimeter wavelengths as well.[24]

Stripes of zebra patterns in the microwave range with separate spike-like pulses of millisecond duration (<8 ms) are related to magnetic reconnection by either whistler wave model[25] or double plasma resonance.[26]

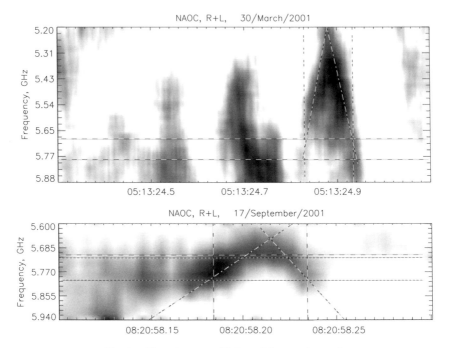

Fig. 3.   The microwave U-shaped bursts observed.

In the latter case, the periodic injection of electron beams into coronal arcade may cause the pulsed spikes in the zebra stripes. The observed quasi-periodic pulsations with about 30 ms period have a peculiar feature of oscillation near a steady state, as shown in Fig. 4, probably resulting from relaxation oscillations, which modulate the electron cyclotron maser emission that forms the zebra stripes during the process of wave–particle interactions.[27]

The zebra patterns at highest frequency regime have been observed.[28] On 13 December 2006 at 03:58 UT another zebra pattern burst in 5.2–5.8 GHz was recorded, and the sub-second spiky zebras were also observed in the 2.6–3.8 GHz, as shown in Figs. 5 and 6.

2.4. Microwave spikes, with absorptive and drifting features

Spikes are generally considered to be closely related to the primary energy release processes,[29,30] though one event shows that they are related to

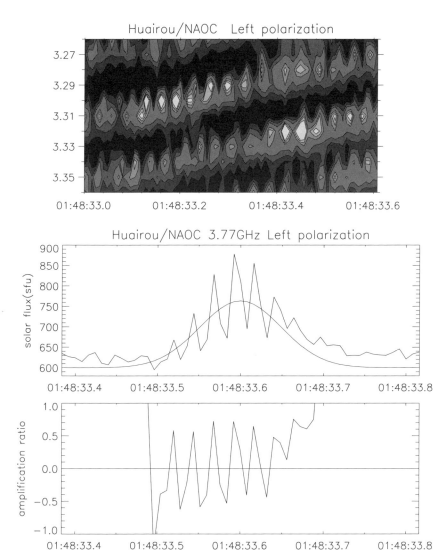

Fig. 4. The microwave zebra pattern bursts with super-fine structures on 21 April 2001 (top). The time profile of one single stripe at 3.77 GHz fitted with a Gaussian function (middle). The ratio is with respect to the Gaussian central line (bottom).[27]

secondary process.[14] A cluster of 99 microwave ms spikes was observed at the highest frequency with center frequency 5.87 GHz and average bandwidth 24.5 MHz, or 0.4%. Duration of each spike is <10 ms.[31] Figure 7 shows the drifting and absorptive spikes.

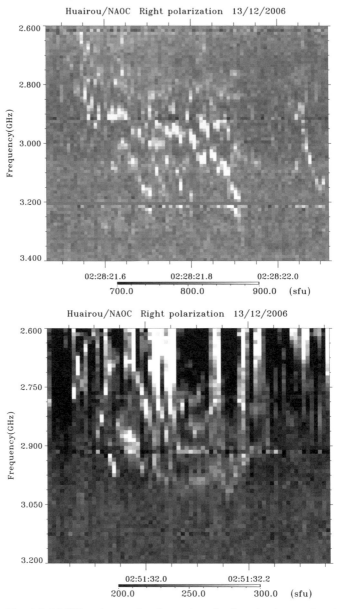

Fig. 5. The 2.6–3.8 GHz sub-second spiky stripes (top) and sub-second stripes with pulsations (bottom) for 13 December 2006 event.

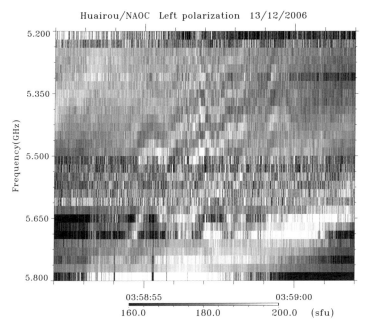

Fig. 6.   The zebra pattern bursts in 5.2–5.8 GHz for 13 December 2006 event.

2.5. *Pulsations and drifting pulsation structures*

Pulsations are also considered to be closely related to the primary energy release processes.[14] The slowly negatively drifting pulsation structure (DPS) was interpreted as the signature of the dynamic magnetic reconnection.[32] Assuming that pulsations are due to the modulation of the tearing mode oscillations of the current-carrying coronal loops, the pulsations in the 13 December 2006 event were analyzed, and the calculations are consistent with the observations.[33]

A statistical study has been carried out and 52 DPS events (51 in 1–2 G, 1 in 2.6–3.8 G) were found in the Huairou/NAOC data set.[34] In the 52 radio burst events, 32 events are long duration ($\geq$10 min) events (LDE). They cover the frequency bandwidth of $0.2 \gtrsim 1$ GHz with global drift rates $-1.6 - -26$ MHz/s. In 32 LDEs, 28 events (88%) are associated with CMEs (27 cases) or ejection (1 case). In 20 nonLDEs, there was no CMEs association below 3 min, but only one weak white light jet. Between 3 and 10 min, there are 17 radio events, in which only 5 events (26%) are accompanied by CMEs, and 8 events are accompanied by weak white light jet. This indicated that the LDEs are more likely to be associated with

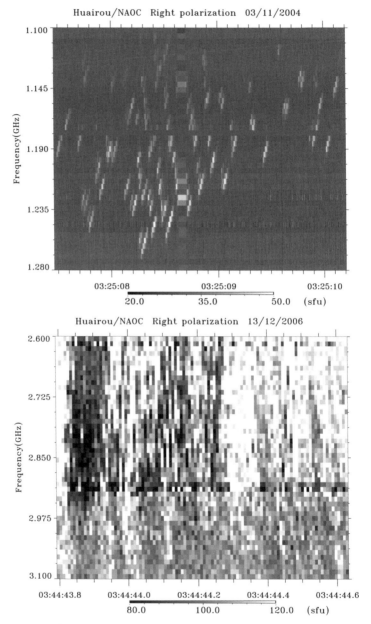

Fig. 7. The drifting spikes around 1.2 GHz on 3 November 2004 (top) and absorptive spike bursts around 3 GHz on 13 December 2006 (bottom).

CMEs. All events are accompanied by GOES SXR flare (10 X-class, 24 M-class, 17 C-class, 1 B-class); 43 (83%) events observed Hα flares (no patrol for nine events); 42 events had EIT observation; 14 events had type-II radio bursts.

3. Summary

In summary, a rich variety of radio FSs at microwave frequencies are presented. Some parameters such as the magnetic field B, the electron density n, etc., from existing models are derived. They should impose constraints on emission mechanisms in the low corona as well as the primary energy release sites. Efforts are needed to understand the physical nature of the observed phenomena, e.g. to build a new instrument, capable of true imaging spectroscopy, with high temporal, spatial, and spectral resolutions — CSRH[35] or FASR.[36]

Acknowledgments

The work was supported by MOST grant (2006CB806301), NSFC grants (10778605, 10333030). NAOC students, J. Huang, B. Chen, and colleagues, Y. Liu, S. Wang, B. Tan, and C. Tan, are acknowledged for their helps.

References

1. T. S. Bastian, A. O. Benz and D. E. Gary, *Ann. Rev. Astron. Astrophys.* **36** (1998) 131.
2. D. E. Gary and C. U. Keller, *Solar and Space Weather Radiophysics* (Kluwer, Dordrecht, 2004).
3. M. J. Aschwanden, *Physics of the Solar Corona* (Springer-Verlag, Berlin, 2004).
4. H. Alvarez and F. T. Haddock, *Solar Phys.* **30** (1973) 15.
5. Q. Fu *et al.*, *Solar Phys.* **222** (2004) 167.
6. G. L. Tarnstrom and K. W. Philip, *Astron. Astrophys.* **17** (1972) 267.
7. H. Isliker and A. O. Benz, *Astron. Astrophys. Suppl.* **104** (1994) 145.
8. K. Jiricka *et al.*, *Astron. Astrophys.* **375** (2001) 243.
9. Y. Yan *et al.*, *IAU Symp.* **226** (2005) 101.
10. V. L. Ginzburg and V. V. Zhelezniakov, *IAU Symp.* **9** (1959) 574.
11. D. Melrose, in *Solar Radiophysics* (Cambridge University Press, Cambridge, 1985), p. 177.
12. A. O. Benz, *Plasma Astrophysics* (Kluwer, Dordrecht, 1993).

13. B. Li, P. A. Robinson and I. H. Cairns, *Phys. Rev. Lett.* **96** (2006) 145005.
14. A. O. Benz, in *Solar and Space Weather Radiophysics*, eds. D. E. Gary and C. U. Keller (Kluwer, Dordrecht, 2004).
15. R. X. Xie *et al.*, *Solar Phys.* **197** (2000) 375.
16. G. Huang *et al.*, *Astrophys. Space Sci.* **259** (1998) 317.
17. Z. Ning, Q. Fu and Q. Lu, *Solar. Phys.* **194** (2000) 137.
18. A. T. Altyntsev *et al.*, *Astron. Astrophys.* **411** (2003) 263.
19. O. Elgaroy, *Nature* **184** (1959) 887.
20. J. Kuijpers, PhD thesis, Utrecht University (1975).
21. A. Krüger, *Introduction to Solar Radio Astronomy and Radio Physics* (D. Reidel Publishing Co., Dordrecht, 1979).
22. C. Slottje, *Atlas of Fine Structures of Dynamic Spectra of Solar Type IV-dm and Some Type II Bursts* (Utrecht University, 1981).
23. G. P. Chernov, *Space Sci. Rev.* **127** (2006) 195.
24. G. P. Chernov *et al.*, *Chin. J. Astron. Astrophys.* **1** (2001) 525.
25. G. P. Chernov *et al.*, *Solar Phys.* **237** (2006) 397.
26. A. A. Kuznetsov and Y. T. Tsap, *Solar Phys.* **241** (2007) 127.
27. B. Chen and Y. Yan, *Solar Phys.* **246** (2007) 431.
28. A. T. Altyntsev *et al.*, *Astron. Astrophys.* **431** (2005) 1037.
29. C. S. Wu and L. C. Lee, *Astrophys. J.* **230** (1979) 621.
30. D. Melrose, *Instabilities in Space and Laboratory Plasmas* (Cambridge University Press, Cambridge, 1986).
31. M. Wang *et al.*, *Solar Phys.* **223** (1999) 201.
32. M. Kliem, M. Karlicky and A. O. Benz, *Astron. Astrophys.* **360** (2000) 715.
33. B. L. Tan *et al.*, *Astrophys. J.* **671** (2007) 964.
34. C. M. Tan, Y. H. Yan and Y. Y. Liu, *Adv. Space Res.* **41** (2008) 969.
35. Y. H. Yan, J. Zhang and G. L. Huang, *Proceedings of the 2004 Asia-Pacific Radio Science Conference* (IEEE Press, New York, 2004), p. 391.
36. T. S. Bastian, *Proceedings of SPIE* **4853** (2003) 98.

Advances in Geosciences
Vol. 14: Solar Terrestrial (2007)
Eds. Marc Duldig *et al.*
© World Scientific Publishing Company

START FREQUENCIES AND ONSET POSITIONS OF
TYPE II RADIO BURSTS IN SOLAR ERUPTIONS

JUN LIN

National Astronomical Observatories of China/Yunnan Astronomical Observatory
Chinese Academy of Sciences, Kunming, Yunnan 650011, P. R. China

Harvard-Smithsonian Center for Astrophysics, 60 Garden Street
Cambridge, MA 02138, USA
jlin@ynao.ac.cn
jlin@cfa.harvard.edu

In a major solar eruptive process, a huge amount of magnetized plasma and energy may be flowing into the outermost corona and interplanetary space at considerable speeds, say a few times 10^3 km s^{-1}. As the bulk of magnetized plasma (coronal mass ejection, namely CME) goes through the corona at such high speed, it is quite likely to invoke and drive a fast-mode shock ahead of it like a piston as its speed exceeds the local magneto-acoustic speed (or the Alfvén speed in the case of the force-free environment like the solar corona). Since a CME has to take off from the still state after the eruption commences, it takes a while before the CME speed exceeds the local Alfvén speed, if it eventually can. So, the lower limit of the altitude exists where the CME-driven shock forms, and thus the type II burst occurs. According to the catastrophe model of solar eruptions, we are able to determine the onset positions of the CME-driven shock in the environment of the solar corona for typical eruptive processes. It is found that the results depend in an apparent way on three parameters: magnetic field, plasma density, and the rate of magnetic reconnection in the current sheet. Comparison with observations indicates good agreement.

1. Introduction

There are two competing classes of mechanisms for invoking the type II radio bursts — those that assume a coronal mass ejection (CME) driven fast shock and those that can produce the so-called blast wave in flare regions. In the former case, fast CMEs may propagate through the corona at speed exceeding the local magneto-acoustic speed, or the Alfvén speed in the force-free environment, exciting the fast mode shock in front of it and invoking the type II radio burst. The long duration gradual solar energetic particle events (see the Introduction of Ref. 1 as a brief but comprehensive review on this issue), the type II radio bursts observed during major eruptions,[2–5] as well as the broadening and the Doppler dimming shown by the spectral profiles of OVI and Lyα lines[6,7] constitute convincing evidences of the passage and the propagation of the CME-driven shocks. Statistical study by Mancuso and Raymond[8] further suggests that all type II radio bursts might be piston-driven, originating at the top or flanks of CMEs, especially when proper geometric effects are taken into account.

An alternative exciter of the type II radio burst is the so-called blast wave which is believed to originate in proximity of the solar flare. This scenario is obviously based on the idea that the flare takes place explosively like a bomb expelling the nearby material in every direction. Unlike the CME-driven shock, to our knowledge, there is not a rigorous theory describing the dynamical properties of the driver of such a wave. The earliest records of the CME/flare-related blast wave in the literature can be found in works by Wagner,[9] Wagner and MacQueen,[10] and Gary et al.[11] In these works, several events that gave rise to type II radio bursts were observed simultaneously by both the HAO Coronagraph/Polarimeter (C/P) on board the Solar Maximum Mission spacecraft and the CSIRO metric wavelength radioheliograph. Composite of C/P images and CSIRO images displayed positions of the CME bubbles and the radio burst excitation regions, and showed that the onset positions of type II emission appeared lower than those of CME bubbles (see Fig. 1, which is taken from Wagner and MacQueen[10]). Wagner and MacQueen[10] thus suggested that the shock (or blast) wave that served to excite type II bursts arosed from the impulsive flare input.

Having carefully re-investigated their data and images (Fig. 1), however, we noticed that the onset positions of the type II emission were not necessarily located behind the CME bubbles. The records of timings for each object shown in Fig. 1 indicate that although CME fronts appeared at higher altitudes than the onset positions of the type II radio burst emissions, all the front edges of CME bubbles were observed later than the onset of type II emissions (e.g., see also Fig. 8 of Ref. 9). This suggests that the CME front was quite likely to be located instantly at the onset position of the type II bursts, and then propagated to the higher altitudes. Another point relevant to this issue is also noteworthy: no correction for projection effects of the images used in the above works was applied. Considering the lower resolution in early works, the above conclusion on the type II burst exciters may not be correct.

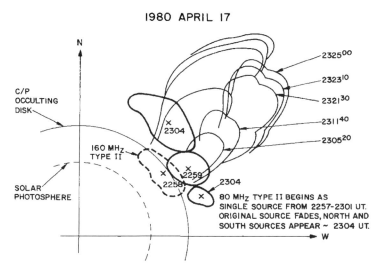

Fig. 1. Composite of Culgoora radioheliograms (bold curves) and C/P white-light images (thin curves) for the 17 April 1980 event (from Ref. 10). No correction for the projection effects was applied.

In the next section we discuss in detail the disadvantage of the blast wave scenario. In Sec. 3, we present the results for the onset positions of and the corresponding start frequencies of type II radio bursts. We summarize this work in Sec. 4.

2. Difficulties in Producing Blast Waves in Flare Regions

The following observational facts make the blast wave origin of the type II burst very questionable. First of all, the traditional flare is the response of the lower atmosphere of the Sun to the energy release (or conversion) occurring in the corona. Such response may perturb the nearby material and propagate to the distance as either a Moreton wave,[12] which is a chromospheric response to the propagation of the coronal fast mode wave (the pressure increase related to a simple-wave or a shock passage pushes the chromospheric plasma downwards at a speed of a few km s^{-1}, which is followed by relaxation, i.e., upward motion after the shock passage), or an EIT wave in the corona.[13,14] But none of these motions involve significant material transportations,[15-17] so, they are unlikely to invoke any kind of shock when propagating.

Second, a typical flare consists of flare ribbons on the disk and flare loops in the corona, which generally trace the local magnetic field configurations. As the flare is in progress, both ribbons and loops manifest outward expansions at speeds that may reach up to $150 \, \mathrm{km\,s^{-1}}$ at the initial stage, and then decreases quickly to a few $\mathrm{km\,s^{-1}}$ in a couple of hours.[18,19] Obviously, the flare loop and ribbon expansion speeds are too small to excite a shock in the media, and furthermore, the motions of flare loops and ribbons are not caused by mass motions of the plasma, but rather by the continual propagation of an energy source (namely the magnetic reconnection site) onto new field lines.[20,21]

Third, observations of type II and type III bursts indicate that type III bursts may appear over a large range of altitudes, from the very low corona (corresponding to the frequency of GHz) to interplanetary space (to the frequency of kHz), and type II bursts do not behave the same way: the reported highest frequencies of the fundamental components of type II bursts usually do not exceed a few hundred MHz.[2,22−27] This actually determines the lowest altitude where type II emission could be excited. On the basis of the empirical model of the plasma density in the corona given by Sittler and Guhathakurta,[28] the frequency of $200 \, \mathrm{MHz}$ corresponds to the altitude of $0.37 \, R_{\odot}$ or $2.6 \times 10^{5} \, \mathrm{km}$ from the solar surface, which is close to the maximum heights where the post-flare loop system could reach near the end of a typical flare process.[19,29]

Fourth, it has been known for decades that about 50% of type II bursts are preceded by a group of type III bursts by several minutes (e.g., Ref. 25, p. 339). Recently, such type III bursts were interpreted as precursors to the associated type II bursts, and their overall frequency envelope (or lane) drifts at a rate close to that of the associated type II bursts.[30,31] This kind of type III bursts usually takes place in the impulsive phase of the flare, and is located between the site of the impulsive energy release and the associated type II bursts.[24] This makes sense if we consult Figs. 2(a) and 2(b), and Fig. 1 of Lin *et al.*,[32] in which a reconnecting current sheet is located between the solar flare and the associated CME, the type II burst is invoked by the CME-driven shock and the beam of energetic particles accelerated by the electric field in the current sheet are responsible for the type III burst.

3. Onset of Type II Radio Burst by CME-Driven Shocks

Discussions in the previous section suggest that the blast (shock) wave which is believed to be produced in the vicinity of the solar flare may not

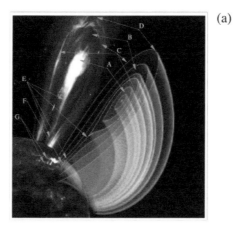

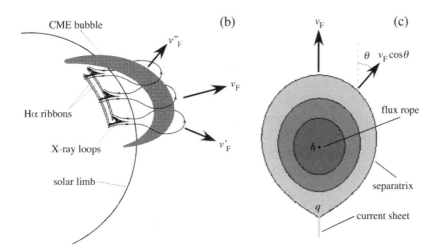

Fig. 2. Morphological features of the typical CME seen from different viewpoints: (a) Composite illustration of a generic CME/flare system and the multi-part structure from LASCO (left image) with the magnetic field structure of the Lin and Forbes[33] model (overlay at right). The LASCO image has been laterally offset for clarity. The CME/flare system comprises: (A) the bright core, (B) the dark void, (C) the outward-moving leading edge, (D) the shock front, (E) the current sheet, (F) the flare loops in the low corona, and (G) their foot points which form the flare ribbons on the solar surface. (From Ref. 34 and courtesy of A. Panasyuk.) (b) Sketch of key components summarized from (a) with a few mathematical notations used in the text. (Original template courtesy of T. G. Forbes.) (c) Schematic descriptions, together with mathematical notes, of the separatrix (CME) bubble in the eruptive process. (From Ref. 35.)

exist in reality, and that the CME-driven shock instead plays an important role in exciting the type II radio bursts. As one of the manifestations of a typical solar eruption, on the other hand, a CME has to take off from a stationary magnetic field at the beginning. So, a CME-driven shock cannot be produced until the CME velocity exceeds the local magneto-acoustic speed. Usually, shocks are expected to form when the CME front speed v_F exceeds the local fast mode speed V_f in the corona, where

$$V_f^2 = V_A^2 \left[\frac{1 + \beta}{2} + \frac{\sqrt{(1 + \beta)^2 - 4\beta \cos^2 \alpha}}{2} \right],$$ (1)

and V_A is the local Alfvén speed, $\beta = (V_s/V_A)^2$ is a factor of 1.2 smaller than the standardly defined plasma beta (V_s is the local sound speed), and α is the angle between the magnetic field and the propagation direction of the shock.[34] For the force-free or the low-beta corona ($\beta \ll 1$), V_f is identified with V_A, which depends on both the local magnetic field and the plasma density and declines with increasing distance after peaking in the low corona.[29] Therefore, as the CME accelerated and as V_A drops, it is likely that the CME can drive a shock when the leading edge speed v_F exceeds the local value of V_f. In the coronal environment, V_f can be well approximated by V_A at the same location.

On the basis of the catastrophe model, we are able to calculate v_F and V_A for the given background field strength σ in units of 100 G, the plasma density distribution in the corona $f(y)$ in units of 10^{10} cm^{-3}, and the rate of magnetic reconnection in the current sheet M_A. Usually, $f(y)$ possesses an empirical form provided by Sittler and Guhathakurta[28] (see also Refs. 29 and 32). With v_F and V_A being known, we are able to locate the onset positions of type II radio bursts by comparing v_F and V_A.

Figure 3 displays four panels plotting v_F and V_A against height (lower x-ticks) and time (upper x-ticks) for various σ and M_A. Arranging x labels and legends this way helps us locate and time the onset of radio bursts easily. Because neither h nor y_B is a linear function of time, especially at the early stage of the eruption, the distribution of the upper x-ticks (for time) is not uniform as we set the lower ones (for altitude) uniformly. Plots in these panels indicate that the positions where v_F surpasses V_A and thus the type II burst is invoked are governed by both σ and M_A in an apparent way.

In Figs. 3(a) and 3(d), the corresponding plasma frequency f_{pe} at the front point of the CME (see point B in Fig. 1 of Ref. 32), which is given by

$$f_{pe}(y) = 0.898408 \sqrt{f(y)}|_{y=y_B} \quad \text{(GHz)}$$ (2)

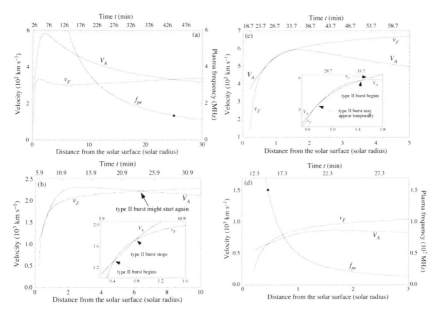

Fig. 3. Variations of the CME front velocities v_F, and the local Alfvén speed V_A versus height and time for different background fields and rates of magnetic reconnection: (a) $M_A = 0.05$ and $\sigma = 0.5$, (b) $M_A = 0.05$ and $\sigma = 2$, (c) $M_A = 0.1$ and $\sigma = 0.69$, and (d) $M_A = 0.1$ and $\sigma = 1$. The corresponding plasma frequencies f_{pe} at the CME front are also plotted in (a) and (d) to demonstrate how f_{pe} varies with altitude. The insets in (b) and (c) show more details of the critical regions, and use the same units.

is also included. Frame ticks for the y-axis in these two panels are arranged such that the left labels and legends are for velocities and the right ones are for the corresponding f_{pe}. Two solid circles on the f_{pe} curves indicate the position and the time at which v_F exceeds V_A. So, the segments of f_{pe} curves to the right of solid circles should depict the main feature of the fundamental lane of type II bursts usually seen in the dynamic spectra of radio emissions.

In Figs. 3(b) and 3(c), we notice that the behavior of v_F compared to V_A becomes more complex than those in the other panels. Two curves meet at two or even three places, which implies richer manifestations of the type II bursts in the eruption. The inset in each panel is used to show more details of v_F and V_A curves in the critical region. Such a complex situation results from the fact that v_F depends on σ, $f(y)$, and M_A, but V_A is governed by the first two parameters only, so they vary with altitude in different fashions. This means that v_F may surpass V_A at some places, then is surpassed by V_A at higher altitudes, and eventually surpasses V_A at much higher altitudes. When this case occurs in reality, we could expect observations of type II radio bursts switching on at high frequency (low altitude), switching off after a while, and then switching on again at low frequency as the associated CME becomes far distant.

As the CME gets far enough from the Sun, on the other hand, deviation of the environment from force-free may mean that v_F needs to surpass V_f given in Eq. (1), instead of V_A, to invoke the fast shock. In this case, even if v_F could exceed V_A, it might not be able to surpass V_f to produce type II bursts. Furthermore, observations indicate that fast CMEs usually underwent deceleration during propagation, which increases the difficulty for v_F to exceed V_f. These facts imply that fast CMEs ($v_F \geq 800\,\mathrm{km\ s^{-1}}$) in some circumstances may not necessarily be able to produce type II radio bursts. Such CMEs are usually known as the radio-quiet fast CMEs, which have been reported several times (Refs. 9, 10 and 39; and references therein).

4. Conclusions

On the basis of the catastrophe model, we investigated the onset positions and the corresponding start frequencies of type II radio bursts occurring in solar eruptions. We compared a couple of possible mechanisms that may account for the type II radio burst, and realized that the fast mode shock driven by CME is most likely to ignite the burst. We found that the onset of the type II burst does not take place at any altitude in the solar atmosphere. Instead, the type II burst never starts at frequency (altitude) higher (lower) than a certain value depending on the magnetic field and plasma background where the eruption occurs, as well as the rate of magnetic reconnection in the same process. Typically, the highest value of the start frequency observed is around 200 MHz, corresponding to the altitude of 0.2 solar radius from the surface. The functional behavior of the CME speed compared to that of the Alfvén speed in the corona determines that even fast ($> 800\,\mathrm{km\,s^{-1}}$) CMEs may not be able to produce type II radio bursts, which is in good agreement with observations.[39] Finally, it is worth noting that the cause of the type II radio burst, especially the metric one, is still an open question. Recent works by Cane and Erickson,[36] Pohjolainen and Lehtinen,[37] and Vršnak *et al.*[38] investigated various eruptive events, and noted that the CME might not be the driver of the shocks that ignite the type II bursts although problems exist with those shocks not being driven by CMEs.

Acknowledgments

The author's work at YNAO was supported by the Ministry of Science and Technology of China under the Program 973 grant 2006CB806303, by the National Natural Science Foundation of China under the grant 40636031 and the grant 10873030, and by the Chinese Academy of Sciences under the grant KJCX2-YW-T04 to YNAO, and he was supported by NASA grant NNX07AL72G when visiting CfA.

References

1. M. A. Lee, *Astrophys. J. Suppl. Ser.* **158** (2005) 38.
2. H. V. Cane, W. C. Erickson and N. P. Prestage, *J. Geophys. Res.* **107**(A10) (2002) 1315.
3. N. Gopalswamy, *Solar and Space Weather Radiophysics: Current Status and Future Developments*, eds. D. E. Gary and C. U. Keller (Kluwer, Boston, 2004), p. 314.
4. N. Gopalswamy, *The Sun and The Heliosphere as an Integrated System*, eds. G. Poletto and S. T. Suess (Kluwer, Boston, 2004), p. 222.
5. A. Vourlidas, *Solar and Space Weather Radiophysics: Current Status and Future Developments*, eds. D. E. Gary and C. U. Keller (Kluwer, Boston, 2004), p. 233.
6. S. Mancuso, J. C. Raymond, J. L. Kohl, Y. Ko, M. Uzzo and R. Wu, *Astron. Astrophys.* **383** (2002) 267.
7. J. C. Raymond *et al.*, *Geophys. Res. Lett.* **27** (2000) 1439.
8. S. Mancuso and J. C. Raymond, *Astron. Astrophys.* **413** (2004) 363.
9. W. J. Wagner, *Adv. Space Res.* **2**(11) (1983) 203.
10. W. J. Wagner and R. M. MacQueen, *Astron. Astrophys.* **120** (1983) 136.
11. D. E. Gary, G. A. Dulk, L. House, R. Illing, C. Sawyer, W. J. Wagner, D. J. McLean and E. Hildner, *Astron. Astrophys.* **134** (1984) 222.
12. G. E. Moreton and H. E. Ramsey, *Pac. Astron. Soc. Pub.* **72** (1960) 357.
13. D. Moses *et al.*, *Solar Phys.* **175** (1997) 571.
14. B. J. Thompson, S. P. Plunkett, J. B. Gurman, J. S. Newmark, O. C. St. Cyr and D. J. Michels, *Geophys. Res. Lett.* **25** (1998) 2465.
15. P. Chen, C. Fang and K. Shibata, *Astrophys. J.* **622** (2005) 1202.
16. A.-C. Donea and C. Lindsey, *Astrophys. J.* **630** (2005) 1168.
17. A. G. Kosovichev and V. V. Zharkova, *Nature* **393** (1998) 317.
18. Z. Švestka, *Solar Flares* (D. Reidel, Boston, 1976).
19. Z. Švestka, *Solar Phys.* **169** (1996) 403.
20. T. G. Forbes and L. W. Acton, *Astrophys. J.* **459** (1996) 330.
21. B. Schmieder, T. G. Forbes, J. M. Malherbe and M. E. Machado, *Astrophys. J.* **317** (1987) 956.
22. G. A. Dulk, Y. LeBlanc, T. S. Bastian and J. Bougeret, *J. Geophys. Res.* **105** (2000) 27343.

23. A. Klassen, H. Aurass, G. Mann and B. J. Thompson, *Astron. Astrophys. Suppl. Ser.* **141** (2000) 357.
24. A. Klassen, S. Pohjolainen and K.-L. Klein, *Solar Phys.* **218** (2003) 197.
25. M. R. Kundu, *Solar Radio Astronomy* (John Wiley & Sons, New York, USA, 1965).
26. G. Mann, A. Klassen, H. Aurass and H.-T. Classen, *Astron. Astrophys.* **400** (2003) 329.
27. E. Y. Zlotnik, A. Klassen, K.-L. Klein, H. Aurass and G. Mann, *Astron. Astrophys.* **331** (1998) 1087.
28. E. C. Sittler Jr. and M. Guhathakurta, *Astrophys. J.* **523** (1993) 812.
29. J. Lin, *Chinese J. Astron. Astrophys.* **2** (2002) 539.
30. A. Klassen, H. Aurass, K.-L. Klein, A. Hofmann and G. Mann, *Astron. Astrophys.* **343** (1999) 287.
31. A. Klassen and S. Pohjolainen, *Solar Variability: From Core to Outer Frontiers*, ed. A. Wilson, ESTEC, Noordwijk, The Netherlands, ESA-SP 506 (2002), p. 307.
32. J. Lin, S. Mancuso and A. Vourlidas, *Astrophys. J.* **649** (2006) 1110.
33. J. Lin and T. G. Forbes, *J. Geophys. Res.* **105** (2000) 2375.
34. J. L. Kohl, G. Noci, S. R. Cranmer and J. C. Raymond, *Astron. Astrophys. Rev.* **13** (2006) 31.
35. J. Lin, J. C. Raymond and A. A. van Ballegooijen, *Astrophys. J.* **602** (2004) 422.
36. H. V. Cane and W. C. Erickson, *Astrophys. J.* **623** (2005) 1180.
37. S. Pohjolainen and N. J. Lehtinen, *Astron. Astrophys.* **449** (2006) 359.
38. B. Vršnak, A. Warmuth, M. Temmer, A. Veronig, J. Magdalenić, A. Hillaris and M. Karlický, *Astron. Astrophys.* **448** (2006) 739.
39. N. Gopalswamy, E. Aguilar-Rodriguez, M. L. Kaiser and R. A. Howard, American Geophysical Union, Fall Meeting 2004, abstract #SH23A-05 (2004).

Advances in Geosciences
Vol. 14: Solar Terrestrial (2007)
Eds. Marc Duldig *et al.*
© World Scientific Publishing Company

RADIO BURSTS IN dm–cm WAVELENGTH RANGE DURING FLARE/CME EVENTS

YIHUA YAN

National Astronomical Observatories
Chinese Academy of Sciences, Beijing 100012, China

The State Key Laboratory of Space Weather
P. O. Box 8701, Beijing 100080, China
yyh@bao.ac.cn

Radio bursts with fine structures in decimetric–centimetric wavelength range are generally believed to manifest the primary energy release process during flare/CME events. Using spectropolarimeters at 1–2 GHz, 2.6–3.8 GHz, and 5.2–7.6 GHz at NAOC/Huairou with very high temporal (1.25–8 ms) and spectral (4–20 MHz) resolution, the zebra patterns, spikes, and new types of radio-fine structures with mixed frequency drift features are observed during several significant flare/CME events. The occurrence of radio-fine structures during the impulsive phase of flares and/or CME initiations are presented, which may be connected to the magnetic reconnection processes.

1. Introduction

Radio-emission generated by thermal and nonthermal electrons, provides important diagnostics in addition to EUV, SXR, HXR, and γ-rays (e.g., Aschwanden[1]). Frequencies in $\sim 10^2$ MHz – <10 GHz correspond to source densities of a few 10^8–10^{11} cm^{-3}, where the primary energy release of flares should take place.[2,3] Radio-fine structures (FSs) such as spikes, zebra patterns, and pulsations are generally considered to be closely related to the primary energy release processes if they occur in the impulsive flare phase.[1-7]

The stripes of the zebra patterns in the microwave range with separate spike-like pulses of millisecond duration (<8 ms) are related to magnetic reconnection above flaring regions by either the whistler wave model[7] or the mechanism of double plasma resonance (DPR).[8] In the latter case, the periodic injection of electron beams into the coronal arcade may cause the pulsed spikes in the zebra stripes. Pulsations are also considered to be closely related to the primary energy release processes.[3]

The slowly negatively drifting pulsation structure (DPS) was interpreted as the signature of the dynamic magnetic reconnection.[6]

2. Radio Bursts Associated with Flare/CME Processes

The Chinese Solar Radio Broadband Fast Dynamic Spectrograph was developed with very high temporal and spectral resolution.[9] In the following, we discuss the observed radio FSs that may be associated with the magnetic reconnection process in the solar corona.

2.1. *Statistic studies*

The occurrence of solar radio burst FSs in the 1–7.6 GHz range associated with CME events was statistically studied with 91 radio burst events selected during 1997–2003.[10] These events were all associated with GOES X-ray flares. There are 38 among 91 events (41.7%) containing FSs (or FS groups regardless of their types). As listed in Table 1, most radio FSs occurred during the rising phase of flare/CMEs, and the occurrence of FSs decreases as frequency increases. Table 2 shows that most events were in a time sequence of radio burst → SXR flare → CME onset (65%).

As mentioned above, the DPS is interpreted as the signature of the dynamic magnetic reconnection and 52 DPS events (51 in the 1–2 GHz range and 1 in the 2.6–3.8 GHz range) during the same period 1997–2003

Table 1.   Occurrence of Huairou radio FSs during flare processes.

Frequency range	Rising phase	Flare peak	Decay phase
1–2 GHz	19	15	4
2.6–3.8 GHz	14	3	3
5.2–7.6 GHz	5	2	1

Table 2.   Time sequence of the radio burst, flare maximum and CME onset.

Time sequence:	R–F–C	R–C–F	F–R–C	F–C–R	C–R–F	C–F–R
Event number:	59	8	12	2	7	3
	65%	9%	13%	2%	8%	3%

Note: The time sequence is represented by "R–F–C," etc., indicating firstly radio bursts occurred, then flare maximum, and finally CME onset, etc.

were found in the data set at Huairou Station of the National Astronomical Observatories of China (Huairou/NAOC).[11] It should be noted that they are not a subset of the above 91 events. There are 32 out of 52 events that are Long Duration ($\geq$10 min) Events (LDE). Of the 32 LDEs, 28 events (88%) are associated with CMEs (27) or ejection (1). In the remaining 20 non-LDEs, for radio bursts with duration less than 3 min no CMEs are found except for 1 weak white light jet. There are 17 radio events with duration 3–10 min, and 5 events are found accompanied by CMEs, whereas the other 8 events are accompanied by weak white light jet. All events are accompanied by GOES SXR flare (10 X-class, 24 M-class, 17 C-class, 1 B-class); 43 events had associated Hα flares (no patrol for 9 events); and 14 events had associated type-II radio bursts.

For the 32 DPS events which were associated with CMEs, three CMEs had speeds of 400–500 km s^{-1}, 16 CMEs had speeds of 501–1000 km s^{-1}, and 13 CMEs had speeds greater than 1000 km s^{-1}.[11] It seems that DPSs are more likely to associate with fast CMEs.

2.2. *Case studies*

2.2.1. *The 1 December 2004 event*

There are three groups of radio FSs during the impulsive phase of this flare that occurred from 07:00 UT–07:40 UT in the 1.100 GHz–1.340 GHz band denoted by N1, Z2, Z3.[12] N1 has several emission lines with mixed fast- and slow-frequency drift rate, as shown in Fig. 1, which may reflect the conditions of a flare loop and fast flows out from a reconnection site; Z2 and Z3 are zebra patterns. When N1 occurred, the HXR changed from a single source to a double source. The fast-positive frequency drift (21.8 MHz s^{-1}) may be due to the fast downward flows (438 km s^{-1}) from the reconnection site, whereas the slow-negative drift (-3.8 MHZ s^{-1}) may be due to the slow upward motion of flare loops (76 km s^{-1}).[12] The scenario can be understood by the simulation as shown in Lin *et al.*,[13] where the upward motion of the flux rope (or the top of the current sheet) is about 1000 km s^{-1}, whereas the upward motion of the top of the flare loops is in a range of 10–100 km s^{-1}.

2.2.2. *The 13 December 2006 event*

The flare productive region NOAA 10930 at a disk location of S05 W33 produced an X3.4/4B flare at 02:40 UT (peak time) on 13 December 2006 during the solar minimum. The flare was accompanied by a halo

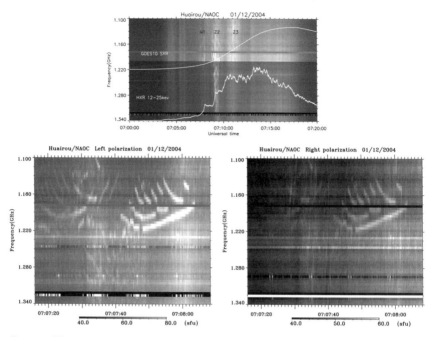

Fig. 1.   The radio spectra and X-ray profiles on 1 December 2004 in the rising phase
of the flare during 07:00–07:20 UT (top) and the spectacular radio FSs of left and right
polarizations (lower panels) during 07:07–07:08 UT.

CME with a projected speed $1500 \, \mathrm{km \, s^{-1}}$. Very rich complicated radio-fine
structures were observed during a long period from 02:20 to 05:10 UT by the
Huairou/NAOC spectropolarimeters with very high temporal (5–8 ms) and
spectral (10–20 MHz) resolutions in 2.6–3.8 GHz and 5.2–7.6 GHz ranges.[9]

During the impulsive phase of the flare in 02:21–02:25 UT groups of
many short period fine structures including spikes, narrow band pulsations,
type-III bursts, fibers, and spiky sub-second zebra-like features appeared
first, and they corresponded to the coronal topological changes as revealed
by Hinode SXR images.[14]

After 02:23:25 UT for nearly 1 min duration, enhanced continuum
emission occurred with a bright RHCP continuum in the frequency range
from less than 2.6 GHz (due to our band limit) to a higher cutoff frequency
up to about 2.9 GHz. Around this continuum there are numerous extensions
that jut out to higher frequencies, as shown in Fig. 2. More than 50
individual sub-second FSs during 02:22–02:25 UT can be registered before a

broad-band continuum enhancement from about 02:23:55 UT. Some spiky sub-second zebra-like super-fine structures and fibers are shown in Fig. 2. These radio observations should be closely related to the primary energy release site and may provide very useful diagnostics in the source region. It is interesting to note that the derived magnetic field scale height would increase by a factor of 2 from the impulsive phase at 02:23 UT to around the flare maximum at 02:43 UT, if the source density scale height is assumed to not change significantly. The estimated coronal field is about 50–170 G in the rising phase of the flare with a source density of about 1×10^{11} cm^{-3}. The field value and plasma density are about 90–200 G and 1.27×10^{11} cm^{-3} around flare maximum.[14]

The occurrence of bi-directional type-IIIs is considered to correspond to the primary acceleration site. The accelerated particles propagating along the magnetic field lines both upward and downward may generate these bi-directional type-III bursts.[1] Drifting bi-directional bursts for about 20 s, or upward motion of the reconnection site, was observed for the first time for the 13 December 2006 flare event, as shown in Figs. 3 and 4. The positive and negative drift rates are about ~ -10 GHz s^{-1} and $+5$ GHz s^{-1} around 03:29 UT and -13 GHz s^{-1} and $+10$ GHz s^{-1} around 03:44 UT with the central frequency around 3 GHz, which indicates the source density of about 10^{11} cm^{-3}.

3. Conclusions

Radio FSs may manifest signatures of the initial phases and provide important diagnostics of coronal parameters during reconnections. For example, for the 13 December 2006 event, the derived magnetic field scale height from the double plasma resonance model would increase by a factor of 2 from the impulsive phase at 02:23 UT to around the flare maximum at 02:43 UT, if the source density scale height is assumed to not change significantly. The estimated coronal field is about 50–170 G in the rising phase of the flare with a source density of about 1×10^{11} cm^{-3}. The field value and plasma density are about 90–200 G and 1.27×10^{11} cm^{-3} around flare maximum. These radio FSs also impose constraints on emission mechanisms related to the reconnection process. To advance the understanding of rich variety of radio FSs and dynamics in solar corona, imaging spectroscopy with high resolutions in all temporal, spatial, and spectral parameters — CSRH[15] or FASR[16] is needed.

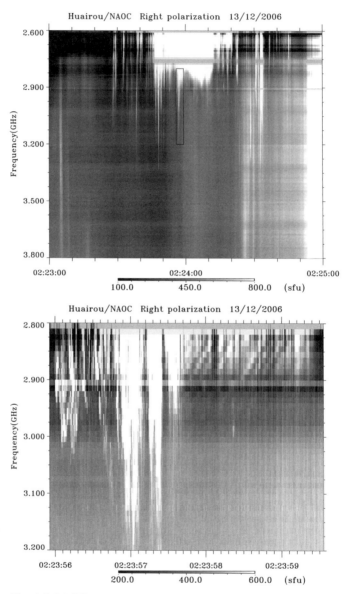

Fig. 2.   The 2.6–3.8 GHz spectrogram in 02:23–02:25 UT during impulsive phase of
the flare corresponding to the coronal topological changes as revealed by Hinode SXR
images.[14] Enlarged FSs including sub-second spikes, type-III bursts, fibers, and spiky
zebra-like features in the box region of the top panel (bottom).

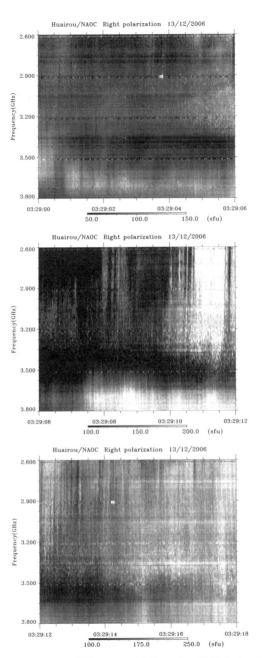

Fig. 3.   The bi-directional type-III bursts on 13 December 2006 during 03:29:00–03:29:18 UT with frequency drift rates of about −10 GHz/s (upward) and +5 GHz/s (downward).

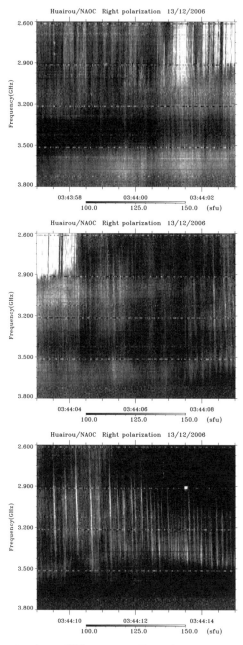

Fig. 4.   The bi-directional type-III bursts on 13 December 2006 during 03:43:57–03:44:15 UT with frequency drift rates of about −13 GHz/s (upward) and +10 GHz/s (downward).

Acknowledgments

The work was supported by MOST grant (2006CB806301), NSFC grants (10778605, 10333030). NAOC students, J. Huang, B. Chen, and colleagues, Y. Liu, S. Wang, B. Tan, and C. Tan, are acknowledged for their helps.

References

1. M. J. Aschwanden, *Physics of the Solar Corona* (Springer-Verlag, Berlin, 2004).
2. T. S. Bastian, A. O. Benz and D. E. Gary, *Ann. Rev. Astron. Astrophys.* **36** (1998) 131.
3. A. O. Benz, in *Solar and Space Weather Radiophysics*, eds. D. E. Gary and C. U. Keller (Kluwer, Dordrecht, 2004).
4. G. D. Holman, D. Eichler and M. R. Kundu, *IAU Symp.* **86** (1980) 457.
5. V. V. Zaitsev and A. V. Stepanov, *Solar Phys.* **88** (1983) 297.
6. M. Kliem, M. Karlicky and A. O. Benz, *Astron. Astrophys.* **360** (2000) 715.
7. G. P. Chernov *et al.*, *Solar Phys.* **237** (2006) 397.
8. A. A. Kuznetsov and Y. T. Tsap, *Solar Phys.* **241** (2007) 127.
9. Q. Fu *et al.*, *Solar Phys.* **222** (2004) 167.
10. Y. Yan *et al.*, *IAU Symp.* **226** (2005) 101.
11. C. M. Tan, Y. H. Yan and Y. Y. Liu, *Adv. Space Res.* **41** (2008) 969.
12. J. Huang, Y. Yan and Y. Liu, *Adv. Space Res.* **39** (2007) 1439.
13. J. Lin, S. Mancuso and A. Vourlidas, *Astrophys. J.* **649** (2006) 1110.
14. Y. Yan, J. Huang, B. Chen and T. Sakurai, *Publ. Astron. Soc. Jpn.* **59** (2007) S815.
15. Y. H. Yan, J. Zhang and G. L. Huang, in *Proceedings of the 2004 Asia-Pacific Radio Science Conference* (IEEE Press, New York, 2004) p. 391.
16. T. S. Bastian, *Proceedings of SPIE* **4853** (2003) 98.

Advances in Geosciences
Vol. 14: Solar Terrestrial (2007)
Eds. Marc Duldig *et al.*
© World Scientific Publishing Company

CORONAL MASS EJECTION RECONSTRUCTIONS FROM INTERPLANETARY SCINTILLATION DATA USING A KINEMATIC MODEL: A BRIEF REVIEW

M. M. BISI*, B. V. JACKSON, P. P. HICK, A. BUFFINGTON
and J. M. CLOVER

Center for Astrophysics and Space Sciences
University of California, San Diego
9500 Gilman Drive #0424, La Jolla, CA 92093-0424, USA
*mmbisi@ucsd.edu
*mario.bisi@gmail.com

Interplanetary scintillation (IPS) observations of multiple sources provide a view of the solar wind at all heliographic latitudes from around 1 AU down to coronagraph fields of view. These are used to study the evolution of the solar wind and solar transients out into interplanetary space, and also the inner-heliospheric response to co-rotating solar structures and coronal mass ejections (CMEs). With colleagues at the Solar Terrestrial Environment Laboratory (STELab), Nagoya University, Japan, we have developed near-real-time access of STELab IPS data for use in space-weather forecasting. We use a three-dimensional (3D) reconstruction technique that obtains perspective views of solar co-rotating plasma and of outward-flowing solar wind crossing our lines of sight from the Earth to the radio sources. This is accomplished by iteratively fitting a kinematic solar wind model to the IPS observations. This 3D modeling technique permits reconstructions of the density and speed structures of CMEs and other interplanetary transients at a relatively coarse resolution. These reconstructions have a 28-day solar-rotation cadence with 10° latitudinal and longitudinal heliographic resolution for a co-rotational model, and a one-day cadence and 20° latitudinal and longitudinal heliographic resolution for a time-dependent model. These resolutions are restricted by the numbers of lines of sight available for the reconstructions. When Solar Mass Ejection Imager (SMEI) Thomson-scattered brightness measurements are used, lines of sight are much greater in number so that density reconstructions can be better resolved. Higher resolutions are also possible when these analyses are applied to Ootacamund IPS data.

1. Introduction

Interplanetary scintillation (IPS) has been used for solar wind, solar wind transient, and inner-heliospheric observations for over 40 years, e.g.

Refs. 1–7. IPS is the rapid variation in signal received by radio antennas on Earth from a compact radio source, arising from scattering by small-scale (~150 km) density inhomogeneities in the solar wind flowing approximately radially outward from the Sun. IPS observations allow the solar wind speed to be inferred over all heliographic latitudes and a wide range of heliocentric distances (dependent upon source strength and observing frequency), e.g. Refs. 1–7. Using the level of scintillation converted to g-level as a proxy, the solar wind density can also be inferred from IPS observations, e.g. Refs. 8 and 9.

As described in detail in Ref. 9, scintillation-level measurements have been available from the Solar Terrestrial Environment Laboratory (STELab)[10] radio antenna at Kiso from 1997 to the present, and more recently from mid-2002 from the STELab radio antenna at Fuji. The New Toyokawa site (see later) will also be used for these measurements. The disturbance factor g is defined by Eq. (1).

$$g = m/\langle m \rangle. \tag{1}$$

$\Delta I/I$ in relation to this equation is the ratio of source intensity variation to measured signal intensity, m is the observed fractional scintillation level, and $\langle m \rangle$ is the modeled mean level of $\Delta I/I$ for the source at the elongation at the time of observation. Scintillation-level measurements from the STELab radio facility analyses are available at a given sky location as an intensity variation of the source signal strength. For each source, data are automatically edited to remove any obvious interference discerned in the daily observations. Further discussion regarding the calculation of and use of g-level as a proxy for density (and also the real-time calculation used for space-weather forecasting) can be found in Refs. 8 and 9.

When two or more radio antennas are used and the separation of the ray-paths in the plane of the sky from source to each telescope lies close to radial (the solar wind flow direction) centered at the Sun, a high degree of correlation between the patterns of scintillation recorded at the two telescopes may be observed, e.g. Ref. 11. The time lag for which maximum cross-correlation occurs (taking into account "plane-of-sky" assumptions) can then be used to estimate the outflow speed of the irregularities producing the scintillation, e.g. Refs. 12 and 13. More sophisticated methods involving the fitting of the observed auto- and cross-correlation spectra with the results from a weak-scattering model, have also

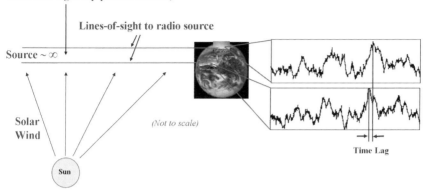

Fig. 1. Figure showing the basic principles of multi-site interplanetary scintillation (IPS) observations through simultaneous observation of a single radio source from multiple (in this case two) antennas. The signal received from a distant, compact source has a variation in amplitude which is directly related to turbulence in the material crossing your line of sight (in this case outflow from the Sun), and thus can be related to variations in density. The example shows similar amplitude variations of signal with a time lag as they pass across the sky from one receiver's line of sight to the other, and are then later used to calculate a measurement of outflow speed. IPS is most sensitive to the point of closest approach to the Sun (P-Point) and to material flowing perpendicular/close-to-perpendicular across the line-of-sight. Figure outline originally courtesy of R.A. Fallows (Aberystwyth University), adapted from Ref. 16.

been adopted for IPS data analyses, e.g. Refs. 7, 14 and 15. Figure 1 shows a picture version of how IPS signals are received using two radio antennas.

The primary sources of data discussed in this chapter were taken from observations made by two different IPS systems. These are the radio arrays of the STELab,[10] University of Nagoya, Japan, and also the Ootacamund (Ooty) Radio Telescope (ORT),[17–19] India; both systems operate at an observing frequency of 327 MHz and both (the new Toyokawa antenna is shown from STELab) are pictured in Fig. 2. STELab typically observes 20–40+ radio sources per day, and Ooty is currently capable of observing up to 1000 radio sources per day.

We use a purely kinematic solar wind model to yield three-dimensional (3D) speed and density reconstructions[8] using a technique that obtains perspective views of solar co-rotating plasma[20] and of outward-flowing solar wind[9] crossing our lines of sight from the Earth to the radio sources, by iteratively fitting our model to the IPS data. We then compare the resulting 3D reconstructions with *in situ* measurements from the

Fig. 2. The new Solar Terrestrial Environment Laboratory (STELab) Toyokawa antenna (left) — currently nearing construction with operation expected in early-mid 2008 (private communication, M. Tokumaru, STELab, 2007), and (right) the Ootacamund (Ooty) Radio Telescope (ORT).
(courtesy of http://www.ncra.tifr.res.in/ NSSS-2008/).

near-Earth Advanced Composition Explorer–Solar Wind Electron, Proton and Alpha Monitor (ACE|SWEPAM),[21,22] and also with "ram" pressure measurements inferred from the Mars Global Surveyor magnetometer,[23] in an orbit around Mars during the time of this set of observations.

Section 2 summarizes the use of 3D speed and density reconstructions from STELab IPS data when compared with "ram" pressure calculations from the Mars Global Surveyor magnetometer, and preliminary results of a "backsided" set of CMEs with their effects seen at Mars. Section 3 summarizes IPS 3D reconstruction work on a flare-related CME event seen by the SOlar and Heliospheric Observatory–Large Angle Spectrometric COronagraph (SOHO|LASCO)[24,25] on 2005/05/13. Section 4 summarizes both speed and density reconstructions of some early-November 2004 geomagnetic storms and discusses the density proxy being improved by using the Solar Mass Ejection Imager[26,27] (SMEI) Thomson-scattered white-light data instead of IPS *g*-level data for density when compared with ACE *in situ* measurements. Section 5 discusses a preliminary analysis and comparison of 3D density reconstructions from both Ooty and STELab IPS data for the early-November 2004 period, and we will give an overall summary in Sec. 6.

2. 3D Reconstructions: Comparison at Mars

An evaluation of both the co-rotating[20] and the time-dependent[9] models using STELab IPS data at the position of Mars is presented in Ref. 6. Both models are used, the first of the two, the co-rotating model, assumes that the heliosphere is unchanging except for outward-flowing solar wind over intervals of one solar rotation. This is where solar rotation provides the primary change of perspective view for each observed location. The second of the two, the time-dependent model, allows time to vary with an interval that is short compared with that of a solar rotation; in this case that of a single day. This short interval imposes the restriction that the reconstructions primarily use the outward motion of the solar wind crossing the lines of sight to give perspective views of each point in space. The 3D reconstruction results using STELab IPS data to date are commensurate with (but also limited by) the observational coverage, temporal and spatial resolution, and also the signal-to-noise level of the observations.

The evaluation in Ref. 6 was carried out through the years 1999–2004 (inclusive) and, since there were no direct measurements of solar wind density or velocity at Mars, solar wind ram pressure measurements derived from the Mars Global Surveyor magnetometer data were used as a solar wind proxy. Equation (2), for transforming the IPS reconstructed solar wind speed and density values extracted at Mars, was formulated in Ref. 6.

$$P = mnv^2 = 2 \times 10^{-6} nv^2. \tag{2}$$

Where P is the derived IPS reconstructed ram pressure at Mars, the effective mass per electron (m) is taken to be 2.0×10^{-24} g; P is in nPa, n is electron number density in $e^- \, cm^{-3}$, and v is speed in $km \, s^{-1}$.

Jackson *et al.*'s[6] 3D IPS reconstructions used two different forms of reconstruction at Mars; a summary of their findings can be seen in Fig. 3. The paper identified 47 independent *in situ* "pressure-pulse" events above 3.5 nPa at Mars (the defined threshold for the investigation) in the Mars Global Surveyor data in time periods from 1999 to 2004 where sufficient STELab IPS data were available. 3D reconstructions using both the co-rotating and time-dependent kinematic models were then calculated from STELab IPS data in terms of both speed and density, and from which a value of pressure was calculated to compare with the Mars *in situ* data using Eq. (2). Time-series of pressure were then plotted from each data set and peaks above 3.5 nPa in the Mars Global Surveyor *in situ* data were compared and cross-correlated in time with corresponding peaks from the reconstructed 3D IPS models.

Even though no "perfect" match was found between the two differing IPS reconstruction models and the inferred *in situ* ram pressure measurements, a very good correlation in time for peak amplitudes was found between each of the models and the recorded data at Mars. Successful

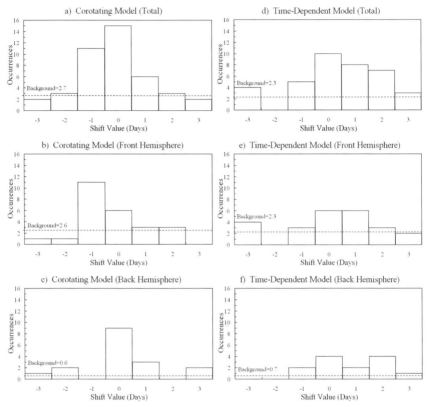

Fig. 3.   Mars IPS reconstruction event summary from Ref. 6. The histograms show
the time lags/leads between peaks in "ram" pressure that were inferred from the Mars
Global Surveyor *in situ* measurements with the corresponding peaks reconstructed in 3D
by both the co-rotating and time-dependent models from the STELab IPS observations.
A positive time shift indicates a lag in the 3D model relative to the Mars *in situ* proxy
and a negative shift indicates a lead in the 3D model relative to the Mars *in situ*
proxy. Part (a) shows the total number of corresponding events; (b) the "front-sided"
corresponding events; and (c) the "back-sided" corresponding events for the co-rotating
model reconstructions. Part (d) shows the total number of corresponding events; (e) the
"front-sided" corresponding events; and (f) the "back-sided" corresponding events for
the time-dependent model reconstructions.

correlation persisted even when Mars was on the opposite hemisphere of
the Sun from the Earth.

   An interesting observation from Ref. 6 that was based on the
assumption that associations of peaks from the Mars *in situ* analyses and
peaks in the IPS modeling analyses are accurate (within a few days), then

the IPS modeling yields solar-wind ram pressures slightly decreased, by about 15%, relative to the pressures observed *in situ* at Mars. This means that the IPS modeling processes produce a lower solar wind speed, a lower solar wind bulk density, or possibly a combination of the two. Moreover, since the Mars Global Surveyor proxy does not account for all terms in the pressure balance, this slightly lower limit on the solar-wind ram pressure indicates that these unaccounted terms must be rather minor contributions to the total Mars magnetospheric solar-wind pressure.[6]

In addition, based on the study carried out by Ref. 6, a peak that appeared just below the 3.5 nPa ram pressure threshold at Mars is thought to be associated with a series of CME events seen in the period 2004/05/30 (30 May 2004) to 2004/06/07 by the SOHO|LASCO instrument which included a back-side Halo CME and several West-limb CME events — these effects were observed at Mars both in the *in situ* data and with the time-dependent IPS reconstruction as seen in Fig. 4. The events are only a "glancing-blow" to Mars, which is likely the cause of the 2–3 day time differential between the two plots at around 9 June 2004, with the reconstruction lagging the arrival time seen *in situ*. This event was first discussed by Refs. 28 and 29.

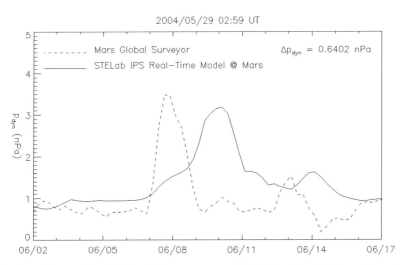

Fig. 4. A time-series of solar-wind ram pressure (nPa) from June 2004 as inferred at Mars from the Mars Global Surveyor magnetometer data (dashed) and also as extracted at the position of Mars from the 3D reconstructed STELab IPS data using the time-dependent model (solid). This is a preliminary analysis adapted from Refs. 28 and 29.

3. 13 May 2005 Flare-Associated CME

The first IPS paper to discuss the 2005/05/13 event was Ref. 30 and this Halo CME has been subsequently discussed by Refs. 31 and 32. A radio-burst resulted from the flare and dimming regions. Both the flare and dimming regions can be seen in the SOHO — Extreme ultra-violet Imaging Telescope (EIT),[33] and circled in Fig. 5. The LASCO images of the CME launch can be seen in Fig. 6.

An interplanetary CME/Magnetic Cloud (ICME/MC) signature was seen by ACE on 2005/05/15. A summary plot from the ACE spacecraft of the solar-wind pressure, magnetic field, radial velocity, and proton density during this day can be seen in Fig. 7. Further details can be found in the captions to the figures.

Again, using g-level as a proxy for density from the STELab IPS observations, we reconstruct the STELab IPS data using the technique described in Ref. 9 and used in Refs. 6, 31, 32 and 34. Note the approximate shape and structure of the ICME as it approaches the Earth on 2005/05/14 as shown in Fig. 8. This is a similar structure to the East of the Sun–Earth line as that seen by SOHO|LASCO in Fig. 6. The timing of the arrival of the event at the Earth from the density reconstructions is approximately consistent with the timing measured by ACE.[28]

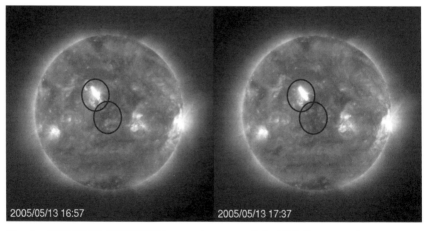

Fig. 5.   2005/05/13 SOHO|EIT images (courtesy of the EIT Consortium) at 16:57 UT (left) and 17:37 UT (right). The active region responsible for this flare/CME (bright area circled) along with associated dimming region (dark area circled) are easily seen. This was a relatively long-lasting active region.

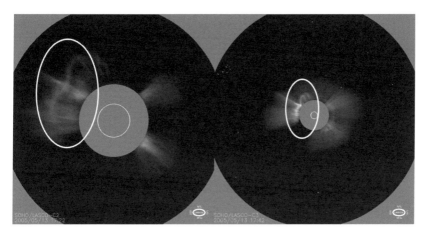

Fig. 6. SOHO|LASCO images of the 2005/05/13 CME, taken from Ref. 32 (courtesy of the LASCO Consortium). The Halo CME pictured at 17:22 UT in LASCO C2 (left) and at 17:42 UT in LASCO C3 (right) with an estimated LASCO speed of 1689 km s^{-1}. CME first C2 appearance was on 2005/05/13 at 17:12:05 UT and the CME first onset at 1R$_\odot$ was on 2005/05/13 at 16:47:34 UT. Notice the double loop-like structure (circled) to the East of the Sun–Earth line in both images.

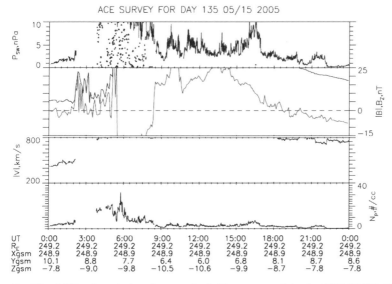

Fig. 7. The ACE solar wind and magnetic field summary data for 2005/05/15, Day of Year (DOY) 135 (adapted from http://pwg.gsfc.nasa.gov/cgi-bin/gif_walk). ACE detected an ICME/MC peak radial velocity on 2005/05/15 of around 1000 km s^{-1}. From the top down: the solar wind pressure in nPa; the absolute magnetic field value (black) and B_z (gray) in nT; the absolute velocity value in km s^{-1} (mostly off the scale); and the proton density in number of protons per cubic centimeter.

2005/05/14 15:00 UT

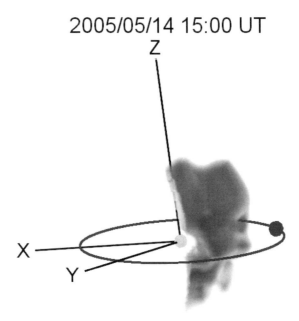

Fig. 8. 3D tomographic reconstruction of the distribution of solar wind density at 15:00 UT on 2005/05/14 as derived from the STELab IPS g-level data using the method described in Ref. 9. All non-associated features of the 2005/05/13 CME (such as behind the Sun relative to the Earth, or in the foreground/background) have been removed. The Sun is represented by the central sphere and the Earth by the outer sphere with its orbital path marked by an ellipse. The view is that of a remote observer East of the Sun–Earth line at a distance of approximately 1.5 AU. The X and Y axes define the Earth's orbital plane and the Z axis is perpendicular to this into the northern heliospheric hemisphere. The lighter the shade, the greater the density in the reconstruction. Density is shown from $8\,e^-\,cm^{-3}$ upward with the square decrease with distance from the Sun removed. Note the double loop-like structure weakly seen here East of the Sun–Earth line which is similar to that seen by LASCO in Fig. 6 but expanded out to over 0.5 AU.

The IPS observations from the radio telescopes of the Multi-Element Radio Linked Interferometer Network (MERLIN)[35] and the European Incoherent SCATter Radar (EISCAT)[36,37] when used to perform extremely long-baseline (ELB) IPS observations, can provide a higher resolution for detecting multiple streams crossing the line of sight and also to the direction of flow, e.g. Ref. 38, of the solar wind across the line of sight, e.g. Refs. 12, 39 and 40. As the baseline for an IPS observation increases, so does the ability to detect and resolve multiple solar wind streams crossing your line of sight at a compromise of reducing the overall level of the signal cross-correlation between the simultaneous IPS signals from two different telescopes of the same radio source.[15] The term ELB IPS has been used since Ref. 39 to describe IPS observations with baselines around 1250 km or greater, e.g. Ref. 12.

A large meridional flow was detected in one of the streams in the EISCAT–MERLIN ELB observations which was first thought to be associated with the 2005/05/13 CME.[40] The 3D density reconstruction was used to constrain the ELB ray-paths by projecting the line of sight from the Earth to the IPS radio source through the 3D volume, and estimating break-points in the form of an angle relative to the Sun to place different structure in different places along the line of sight. From this method of constraining the ELB ray-paths, Breen *et al.*[31] found that there were most likely three different streams detected in the observations which corresponded to three peaks from their weak-scattering tri-modal model[7] used to fit the ELB IPS observations. Previously, Bisi[40] had reported that the large off-radial flow detected was possibly due to the flow of the ICME/MC itself, but by using the tri-modal fit constrained more accurately by the 3D density reconstruction, it was found that the large meridional flow ($\sim$7°–10° pole-ward) is more likely that of the deflected fast solar wind to the solar North of the ICME/MC,[31,32] and not a meridional flow of the ICME/MC itself. The ICME/MC detected by the ELB IPS observations was most likely flowing in a radial direction, although this is not fully determined. It is not clear whether the large pole-ward meridional flow is the direction of the flow of material, or is attributed to a deflection of the magnetic-field North of the ICME/MC. In addition, this 3D tomography was also applied using the time-dependent model for the first time to combined MERLIN/EISCAT/EISCAT Svalbard Radar[41] (ESR) IPS data in Ref. 32.

4. Early-November 2004 Geoeffective CME-Events

This early-November 2004 period was a time of complex activity where multiple CME features (including several Halo CMEs) were seen in both the coronagraph images and their interplanetary counterparts from spacecraft *in situ* plasma and magnetic-field measurements near the Earth. This period included several ICMEs that occurred due to a series of solar eruptions originating from the Sun between 2004/11/04 and 2004/11/08. During this period, there were two ICMEs/MCs which had their magnetic orientations in the opposite direction to one another despite the fact that these events were related to flares coming from above the same active region on the Sun, where that active region's magnetic configuration remained unchanged throughout.[42] A thorough description and discussion of the *in situ* response to these two major geomagnetic storms can be found in Ref. 42.

The Living With a Star (LWS) Coordinated Data-Analysis by Refs. 43 and 44 also includes these early-November 2004 events during their extensive analyses of large geoeffective storms. They defined their large storms as having a *Dst* (disturbance storm time index) ≤ -100 nT for storms occurring between the years 1996 and 2005. They list two possible sources for each of the two large storms defined by the *Dst* criterion with the first storm being on 2004/11/08 and the second on 2004/11/10. The sources for the first storm were seen in SOHO|LASCO C2 on 2004/11/04 at 23:30 UT and at 09:54 UT. The second storm's sources occurred on 2004/11/07 at 16:54 UT and 2004/11/06 at 02:06 UT. Both Refs. 43, 44 and Ref. 42 report that there were multiple interplanetary scintillation signatures caused by each of these two geomagnetic storms.

Also part of the LWS Coordinated Data-Analysis work was carried out by Ref. 34. They show a combination of 3D reconstructions using data from SMEI in terms of Thomson-Scattered white-light brightness as a proxy for density (preliminary analyses), and STELab IPS observations in both *g*-level (as a proxy for density) and speed. They compared reconstructed structures of SMEI density and STELab IPS speed for events during the early-November 2004 period with *in situ* measurements taken by the ACE spacecraft in order to help validate the 3D tomographic reconstruction results. The geomagnetic storms were fairly well reproduced in both the preliminary SMEI density and the IPS speed reconstructions in terms of their timing with LASCO events and with *in situ* comparisons. Figure 9, taken from Ref. 42, shows the STELab IPS density and speed reconstructions on 2004/11/09 at 03:00 UT. This shows the Earth-directed structure seen in 3D as viewed by a remote-observer at around 1.5 AU.

Initial IPS-only 3D reconstructions for these events were good for the speed when compared with ACE measurements, but were not so good for density.[29] IPS speed incorporated into the SMEI reconstructions yielded a better shape for the *in situ* ACE comparisons, and also resulted in a slightly higher correlation of the results as seen in Fig. 10.

The SMEI reconstructions are at a higher temporal and spatial resolution, typically ~ 3 times finer in resolution than those of the IPS reconstructions. SMEI reconstructions are currently limited only by computer analysis considerations and the resulting computation times. This is due to the much more numerous available lines of sight since SMEI is not restricted by the number of bright astronomical radio sources in the sky

2004/11/09 03:00

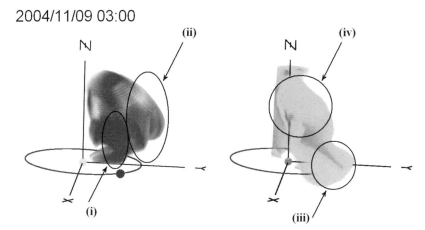

Fig. 9.   3D STELab IPS density (left) and speed (right) tomographic reconstructions, taken from Ref. 42. The reconstructions show the distribution of solar wind density and speed on 2004/11/09 at 03:00 UT. The reconstructions were again carried out using the method described in Ref. 9. All non-associated features have been removed. In both cases, the Sun is represented by the central sphere and the Earth by the outer sphere with its orbital path marked by an ellipse. The view is that of a remote observer partially East of the Sun–Earth line out to a distance of approximately 1.5 AU. The X and Y axes define the Earth's orbital plane and the Z axis is perpendicular to this into the northern heliospheric hemisphere. The lighter the shade, the greater the values of each parameter in the reconstructions. Density is shown from $15\,e^-\,cm^{-3}$ upward to $50\,e^-\,cm^{-3}$, with the square decrease with distance from the Sun removed, and speed is shown from $900\,km\,s^{-1}$ and up. Various points are marked on the figure and are summarized from Ref. 42. (i) Shows the 2004/11/07 event as seen in LASCO C2 at 16:54 UT. (ii) Shows the combination of the two 2004/11/06 events as seen in LASCO C2 at 01:31 UT and 02:06 UT. (iii) Shows a high speed structure engulfing the Earth; this structure which lags the 2004/11/06 events but precedes the 2004/11/07 event is also comparable in speed to that detected by LASCO C2 for the 2004/11/07 event. Finally, (iv) shows high speed solar wind going mainly northward; consistent with the speeds of (iii).

(as is the IPS). To compare with ACE proton density measurements, the present preliminary analysis includes an electron excess due to helium and heavier ions and conversion from SMEI surface-brightness units (analogue-to-digital units, ADUs) to S10 of 0.5 ADU = one S10 was used by the tomography here. An S10 is the intensity of a 10th magnitude star filling one square-degree of sky (see Ref. 45). IPS speed data were incorporated along with the SMEI brightness data (as described by Ref. 46) to improve the global propagation times of SMEI density structures coming out from the Sun in the SMEI reconstructions. The SMEI reconstructions here have bins of 6.7° by 6.7° in latitude and longitude at a 1/2-day temporal cadence. This is described in detail in Ref. 46. The comparison in Fig. 10 shows the

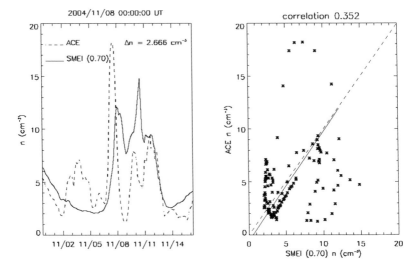

Fig. 10. Comparison plot of SMEI reconstructed density incorporating the IPS speed proxy extracted at the point of the ACE spacecraft for direct-comparison with *in situ* measurements — November 2004 events adapted from Ref. 29. The left plot shows the comparison of the reconstructed density values extracted at the position of the ACE spacecraft from the SMEI Thomson-scattered white-light brightness observations (solid line) and those measured by ACE (dashed line). On the right is a plot of the correlation of these two data sets, the dotted line where a 100% correlation would be found having a one-to-one correspondence. The solid line represents the best-fit of the correlation between the two data sets. Further details are covered in Ref. 34.

ACE data averaged with box-car averaging over a 1/2-day cadence to match that of the SMEI temporal cadence.

The shapes in the reconstructions reproduced around the Earth from the SMEI data show the combination of the several Earth-directed events. These are consistent with the timings of the geoeffective storms described in Ref. 42. The IPS speed data show the fast CME speeds seen in LASCO heading to the North and North-West as well as engulfing the Earth in some high speed wind consistent with what was seen *in situ* at ACE.

5. Preliminary Ooty–STELab 3D Density Reconstruction-Comparisons

Some preliminary analyses using the 3D kinematic time-dependent model have been carried out on the early-November 2004 period with Ooty IPS *g*-level data comparing with STELab *g*-level data in terms of density

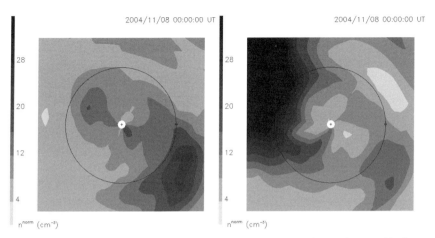

Fig. 11. Figure showing a side-by-side comparison cut in the ecliptic-plane as if looking down from the North pole of the Sun out to 1.5 AU (further at the edges) from STELab (left) and Ooty (right) density reconstructions. Earth's orbit is shown as a thin black circle with the Earth, a small ⊕, indicated on each plot (to the right in each image). The expected r^{-2} density fall-off scaling is used to normalize structures at different radii. Density contours to the left of each image are scaled to 1 AU.

reconstructions alone; not incorporating the IPS speeds at this time (as seen in Figs. 11 and 12). At this preliminary stage, even though the Ooty observations are more numerous than those of STELab, the resolution of the reconstructions was not increased. Two figures, both from 2004/11/08 at 00:00 UT show some similarities and differences between the density reconstructions from each IPS data set.

Figure 11 shows a side-by-side comparison cut in the ecliptic-plane as if looking down from the North pole of the Sun out to 1.5 AU (further at the edges). Figure 12 also shows a side-by-side comparison, this time a cut in the meridional-plane as if looking from 90° East of the Sun–Earth line out to a distance of 1.5 AU from the Sun (again further at the edges). In both figures, the STELab IPS density reconstruction-cut is on the left and the Ooty density reconstruction-cut is on the right. We are unsure if the anti-Earthward directed material reconstructed here only from the Ooty data is real or some kind of artefact from noise in the data propagating through into the reconstruction. The general structure seen to the North and East of the Sun–Earth line is seen in both reconstructions, but to a lesser extent to the East in the Ooty reconstruction. These are just preliminary comparisons at present, and a more-detailed analysis is expected to be undertaken in a forthcoming paper.

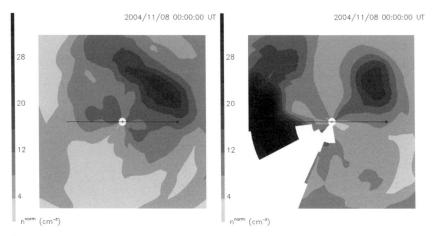

Fig. 12.   Figure showing a side-by-side comparison cut in the meridional-plane as if looking from 90° East of the Sun–Earth line out to a distance of 1.5 AU from the Sun (further at the edges) from STELab (left) and Ooty (right) density reconstructions. Earth's orbit is shown as a thin black line with the Earth, a small ⊕, indicated on each plot (to the right in each image). The expected r^{-2} density fall-off scaling is used to normalize structures at different radii. Density contours to the left of each image are scaled to 1 AU.

6. Summary

This chapter provides a brief summary of the most recent highlights of the 3D tomography reconstruction technique using both the co-rotating and the time-dependent kinematic models, and their applications to various IPS data sets and their extension to employ SMEI data. These include comparisons at Mars, comparisons with near-Earth *in situ* measurements, and also the constraining higher resolution extremely long-baseline IPS observations.

We have summarized the results of the IPS 3D reconstruction techniques in a comparison with *in situ* solar-wind ram-pressure analyses at Mars from the Mars Global Surveyor. This study does not specifically address the forecast capability of this technique at various positions in the inner heliosphere as demonstrated with our near-real-time analyses of the IPS data (http://ips.ucsd.edu/index_ss.html). However, these same modeling techniques provide a forecast of solar-wind conditions at Mars when the IPS arrays are operating, and also at other planets/spacecraft such as Mercury, Venus, Ulysses,[47] and both Solar TErrestrial Relations Observatory (STEREO)[48] spacecraft; thus they have the potential to provide a forecast of solar-wind conditions almost anywhere in the inner heliosphere and sometimes several days in advance for points furthest from the Sun. No spacecraft at Mars currently monitors solar-wind velocity and density regularly. If *in situ* solar-wind monitoring instruments are present on spacecraft near the inner-planets for example, then comparisons with the IPS and/or SMEI 3D reconstructions should become even more relevant and the accuracy improved upon from the study discussed here.

Using the UCSD 3D density reconstructions from STELab IPS data to constrain the more-sensitive ELB observations from EISCAT and MERLIN has the potential to be a very powerful tool.[31,32] It has resulted in our ability to retrieve further information than previously from these very few but highly sensitive observations to detect solar wind directionality and multiple streams along the line of sight. It is hoped that we will be able to use this technique to help constrain and better-fit ELB IPS observations in the future using both the STELab density reconstructions demonstrated here (and the constraining technique fully described in Ref. 31), and using the data from other IPS systems and, of course, from SMEI.

The geoeffective storms discussed here and by Refs. 29, 34, 42–44 are fairly well reproduced both in terms of the IPS speed 3D reconstructions and the preliminary SMEI density 3D reconstructions. They are consistent with the SOHO|LASCO events seen at that time and have been shown to be associated with known *in situ* signatures. The IPS data show the fast CME speeds seen in LASCO heading to the North and North-West as well as those engulfing Earth during the same time period. The structures reproduced around the Earth from SMEI data show a combination of several Earth-directed/near-Earth-directed events. These structures seen in SMEI are consistent with the timing of the geoeffective storms during this period.

The preliminary comparisons between the Ooty and the STELab data are a promising start. Already, without any additional calibration, similar features are seen in both reconstructed data sets. Overall, the Ooty data appear to show enhanced density values compared with the STELab density values when time-series of the two are compared, but we are working on improving this and if necessary, will perform a re-calibration of the kinematic solar wind model to work more accurately with the Ooty data and also aim for higher-resolutions by taking advantage of the more numerous IPS observations. Incorporating the Ooty speed data into the 3D reconstructions will also aid in improving the accuracy of these preliminary reconstructions from data from the Ooty system.

In conclusion, we follow CMEs from near the solar surface outward until they are observed *in situ* near Earth and Mars, and at other points in the inner heliosphere and aim to compare in real-time with both STEREO spacecraft shortly, as is already being done routinely with ACE. These events, reconstructed in 3D in terms of both speed and density, show that the heliospheric response to CMEs is often enormous (from both various IPS data-sets and SMEI observations). We look forward to other (multi-point) *in situ* comparisons such as with Ulysses during its close-pass to the Sun recently in August 2007 and its quadrature earlier in 2007, and other such International Heliophysical Year (IHY) IPS collaborations. As our 3D tomographic models become more sophisticated, possibly incorporating a 3D MHD solar wind model, and multi-point calibrations are realized, we expect the comparisons to improve.

Acknowledgments

The authors acknowledge NSF grant FA8718-04-C-0050, NASA grant NNG05GM58G, and AFOSR grant FA9550-06-1-0107, to the University of California at San Diego (UCSD) for support to work on these analyses. The authors would especially like to thank the group at STELab, Nagoya University (M. Kojima, M. Tokumaru, K. Fujiki, and students) for their continued support, and for making IPS data sets available under the auspices of a joint collaborative agreement between the Center for Astrophysics and Space Sciences (CASS) at UCSD and STELab. Also wish to thank P. K. Manoharan for providing the Ooty (ORT) IPS *g*-level data and the Aberystwyth University IPS group (A. R. Breen and R. A. Fallows) for access to the EISCAT/MERLIN IPS data and results. SMEI was designed and constructed by a team of scientists and engineers from the US Air Force Research Laboratory, the University of California at San Diego, Boston College, Boston University, and the University of Birmingham in the UK. The authors wish to thank the ACE|SWEPAM group for use of the solar wind proton density and velocity measurements used in this chapter for *in situ* comparisons. In addition, the authors would like to thank both the LASCO and EIT consortia for the use of the SOHO|LASCO and SOHO|EIT images in this paper. Thanks also go to D. H. Crider for providing us with the Mars Global Surveyor *in situ* data for these analyses.

References

1. A. Hewish, P. F. Scott and D. Wills, *Nature* **203** (1964) 1214.
2. W. A. Coles and S. Maagoe, *Geophys. Res.* **77** (1972) 5622.
3. W. A. Coles, *Space Sci. Rev.* **72** (1995) 211.
4. R. A. Fallows, A. R. Breen, P. J. Moran, A. Canals and P. J. S. Williams, *Adv. Space Res.* **30** (2002) 437.
5. M. Kojima, A. R. Breen, K. Fujiki, K. Hayashi, T. Ohmi and M. Tokumaru, *J. Geophys. Res. (Space Phys.)* **109** (2004) 4103.
6. B. V. Jackson, J. A. Boyer, P. P. Hick, A. Buffington, M. M. Bisi and D. H. Crider, *Solar Phys.* **241** (2007) 385.
7. M. M. Bisi, R. A. Fallows, A. R. Breen, S. R. Habbal and R. A. Jones, *J. Geophys. Res. (Space Phys.)* **112** (2007) A06101.
8. P. P. Hick and B. V. Jackson, in *Proc. SPIE*, eds. S. Fineschi and M. A. Gummin (2004).
9. B. V. Jackson and P. P. Hick, *Astrophysics and Space Science, Lib.* Vol. 314 (Kluwer Academic Publ., Dordrecht, 2005), pp. 355–386.
10. M. Kojima and T. Kakinuma, *J. Geophys. Res.* **92** (1987) 7269.
11. J. W. Armstrong and W. A. Coles, *J. Geophys. Res.* **77** (1972) 4602.
12. A. R. Breen, R. A. Fallows, M. M. Bisi, P. Thomasson, C. A. Jordan, G. Wannberg and R. A. Jones, *J. Geophys. Res. (Space Phys.)* **111** (2006) 8104.
13. M. M. Bisi, A. R. Breen, R. A. Fallows, G. D. Dorrian, R. A. Jones, G. Wannberg, P. Thomasson and C. Jordan, *EOS Trans. AGU, Fall Meeting Supp. — Abstract SH33A-0399* **87** (2006), p. 52.
14. W. A. Coles, *Astrophys. Space Sci.* **243**(1) (1996) 87.
15. M. Klinglesmith, PhD thesis, University of California, San Diego (UCSD) (1997).
16. M. M. Bisi, A. R. Breen, S. R. Habbal and R. A. Fallows, *"Auld Reekie" MIST/UKSP Joint Meeting*, Edinburgh, Scotland (2004) (Oral presentation).
17. G. Swarup, N. V. G. Sarma, M. N. Joshi, V. K. Kapahi, D. S. Bagri, S. H. Damle, S. Ananthakrishnan, V. Balasubramanian, S. S. Bhave and R. P. Sinha, *Nature* **230** (1971) 185.
18. P. K. Manoharan, M. Kojima, N. Gopalswamy, T. Kondo and Z. Smith, *Astrophys. J.* **530** (2000) 1061.
19. P. K. Manoharan, M. Tokumaru, M. Pick, P. Subramanian, F. M. Ipavich, K. Schenk, M. L. Kaiser, R. P. Lepping and A. Vourlidas, *Astrophys. J.* **559** (2001) 1180.
20. B. V. Jackson, P. P. Hick, M. Kojima and A. Yokobe, *J. Geophys. Res.* **103** (1998) 12049.
21. E. C. Stone, A. M. Frandsen, R. A. Mewaldt, E. R. Christian, D. Margolies, J. F. Ormes and F. Snow, *Space Sci. Rev.* **86** (1998) 1.
22. D. J. McComas, S. J. Bame, P. Barker, W. C. Feldman, J. L. Phillips, P. Riley and J. W. Griffee, *Space Sci. Rev.* **86** (1998) 563.
23. D. H. Crider, D. Vignes, A. M. Krymskii, T. K. Breus, N. F. Ness, D. L. Mitchell, J. A. Slavin and M. H. Acuña, *J. Geophys. Res. (Space Phys.)* **108** (2003) 1461.

24. P. H. Scherer, R. S. Bogart, R. I. Bush, J. T. Hoeksema, A. G. Kosovichev, J. Schou, W. Rosenberg, L. Springer, T. D. Tarbell, A. Title, C. J. Wolfson, I. Zayer and M. E. Team, *Solar Phys.* **162** (1995) 129.

25. G. E. Brueckner, R. A. Howard, M. J. Koomen, C. M. Korendyke, D. J. Michels, J. D. Moses, D. G. Socker, K. P. Dere, P. L. Lamy, A. Llebaria, M. V. Bout, R. Schwenn, G. M. Simnett, D. K. Bedford and C. J. Eyles, *Solar Phys.* **162** (1995) 357.

26. C. J. Eyles, G. M. Simnett, M. P. Cooke, B. V. Jackson, A. Buffington, P. P. Hick, N. R. Waltham, J. M. King, P. A. Anderson and P. E. Holladay, *Solar Phys.* **217** (2003) 319.

27. B. V. Jackson, A. Buffington, P. P. Hick, R. C. Altrock, S. Figueroa, P. E. Holladay, J. C. Johnston, S. W. Kahler, J. B. Mozer, S. Price, R. R. Radick, R. Sagalyn, D. Sinclair, G. M. Simnett, C. J. Eyles, M. P. Cooke, S. J. Tappin, T. Kuchar, D. Mizuno, D. F. Webb, P. A. Anderson, S. L. Keil, R. E. Gold and N. R. Waltham, *Solar Phys.* **225** (2004) 177.

28. M. M. Bisi, B. V. Jackson, P. P. Hick and A. Buffington, *LWS CDAW 2007 Meeting* (2007) (Oral presentation).

29. M. M. Bisi, B. V. Jackson, P. P. Hick and A. Buffington, *AOGS 2007 Meeting* (2007) (Oral presentation).

30. R. A. Jones, A. R. Breen, R. A. Fallows, M. M. Bisi, P. Thomasson, G. Wannberg and C. A. Jordan, *Annales Geophysicae* **24** (2006) 2413.

31. A. R. Breen, R. A. Fallows, M. M. Bisi, R. A. Jones, B. V. Jackson, M. Kojima, G. Dorrian, H. R. Middleton, P. Thomasson and G. Wannberg, *Astrophys. J. Lett.* **683** (2008) L79.

32. M. M. Bisi, B. V. Jackson, R. A. Fallows, A. R. Breen, P. P. Hick, G. Wannberg, P. Thomasson, C. A. Jordan and G. D. Dorrian, *Proceedings SPIE Optical Engineering + Applications 2007 Meeting*, April 2007.

33. J. P. Delaboudiniere, G. E. Artzner, J. Brunaud, A. Gabriel, J. F. Hochedez, F. Millier, X. Y. Song, B. Au, K. P. Dere, R. A. Howard, R. Kreplin, D. J. Michels, J. D. Moses, J. M. Defise, C. Jamar, P. Rochus, J. P. Chauvineau, J. P. Marioge, R. C. Catura, J. R. Lemen, L. Shing, R. A. Stern, J. B. Gurman, W. M. Eupert, A. Maucherat, F. Clette, P. Cugnon and E. L. van Dessel, *Solar Phys.* **162** (1995) 291.

34. M. M. Bisi, B. V. Jackson, P. P. Hick, A. Buffington, D. Odstrcil and J. M. Clover, *J. Geophys. Res. — Geomagnetic Storms of Solar Cycle 23* **113** (2008) A00A11, doi: 10.1029/2008JA01322.

35. P. Thomasson, *Quart. J. Royal Astron. Soc.* **27** (1986) 413.

36. H. Rishbeth and P. J. S. Williams, *Monthly Notices Royal Astron. Soc.* **26** (1985) 478.

37. G. Wannberg, L.-G. Vanhainen, A. Westman, A. R. Breen and P. J. S. Williams, in *Conference Proceedings, Union of Radio Scientists (URSI)* (2002).

38. P. J. Moran, A. R. Breen, C. A. Varley, P. J. S. Williams, W. P. Wilkinson and J. Markkanen, *Annales Geophysicae* **16** (1998) 1259.

39. M. M. Bisi, A. R. Breen, R. A. Fallows, P. Thomasson, R. A. Jones and G. Wannberg, in *ESA SP-592: Solar Wind 11/SOHO 16, Connecting Sun and Heliosphere*, September 2005.

40. M. M. Bisi, PhD thesis, The University of Wales, Aberystwyth (2006).

41. G. Wannberg, I. Wolf, L.-G. Vanhainen, K. Koskenniemi, J. Röttger, M. Postila, J. Markkanen, R. Jacobsen, A. Stenberg, R. Larsen, S. Eliassen, S. Heck and A. Huuskonen, *Radio Sci.* **32** (1997) 2283.

42. L. K. Harra, N. U. Crooker, C. H. Mandrini, L. van Driel-Gesztelyi, S. Dasso, J. Wang, H. Elliott, G. Attrill, B. V. Jackson and M. M. Bisi, *Solar Phys.* **244** (2007) 95, doi: 10.1007/s11207-007-9002-x.

43. J. Zhang, I. G. Richardson, D. F. Webb, N. Gopalswamy, E. Huttunen, J. C. Kasper, N. V. Nitta, W. Poomvises, B. J. Thompson, C.-C. Wu, S. Yashiro and A. N. Zhukov, *J. Geophys. Res.* **112** (2007) 10102.

44. J. Zhang, I. G. Richardson, D. F. Webb, N. Gopalswamy, E. Huttunen, J. C. Kasper, N. V. Nitta, W. Poomvises, B. J. Thompson, C.-C. Wu, S. Yashiro and A. N. Zhukov, *J. Geophys. Res.* **112** (2007) 12103.

45. B. V. Jackson, A. Buffington, P. P. Hick, X. Wang and D. Webb, *J. Geophys. Res. (Space Phys.)* **111** (2006) 4.

46. B. V. Jackson, M. M. Bisi, P. P. Hick, A. Buffington, J. M. Clover and W. Sun, *J. Geophys. Res. — Geomagnetic Storms of Solar Cycle 23* **113** (2008) A00A15, doi: 10.1029/2008JA013224.

47. F. P. Wenzel, R. G. Marsden, D. E. Page and E. J. Smith, *Astronomy and Astrophysics Supplement Series* **92** (1992) 207.

48. M. L. Kaiser, *Adv. Space Res.* **36** (2005) 1483.

Advances in Geosciences
Vol. 14: Solar Terrestrial (2007)
Eds. Marc Duldig *et al.*
© World Scientific Publishing Company

SPECTRUM OF DENSITY FLUCTUATIONS
IN THE SOLAR WIND

V. KRISHAN

Indian Institute of Astrophysics, Bangalore-560034, India

Raman Research Institute, Bangalore-560080, India

vinod@iiap.res.in

The power spectrum of the density fluctuations in the solar wind and its relationship with the spectra of the velocity and the magnetic field fluctuations is investigated. The density fluctuations could behave as a passive scalar and be simply convected by the velocity or the magnetic field fluctuations or they could dynamically participate in the joint production mechanism of all the fluctuations. The spectrum of the density fluctuations can distinguish between these two possibilities. Further, the inclusion of the Hall effect near the ion-inertial scale generates different spectra for the velocity and magnetic fluctuations. Which spectrum would the density fluctuations, behaving as a passive scalar, follow in such a case? The answer leads to the interesting consequence that the electron density fluctuations and the ion density fluctuations have different spectra at spatial scales equal to and smaller than the ion-inertial scale. This result clearly demonstrates the two-fluid picture brought in by the Hall effect.

1. Introduction

The study of the solar wind turbulence is of immence importance on several counts. A distant plasma system that is accessible to direct observations of nonlinear processes such as the wave–particle and wave–wave interactions have an essential bearing on the propagation of the cosmic rays. The coupling of the solar wind and the magnetosphere defines the solar–terrestrial relationship. The turbulence modifies the transport processes.

The reduced spectra of the fluctuations are obtained by averaging over the two directions perpendicular to the solar wind velocity V_s. The spectra are a function of the wavenumber along V_s. The spectral energy distributions of the velocity and the magnetic field fluctuations in the solar wind are now known in a wide frequency range, starting from much below the proton cyclotron frequency (0.1–1 Hz) to hundreds of Hz.

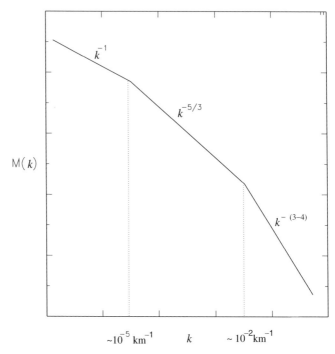

Fig. 1.   Schematic representation of the observed magnetic energy spectrum in the solar wind.

The inferred power spectrum[1] of magnetic fluctuations (Fig. 1) consists of multiple segments — a Kolmogorov-like branch ($\propto k^{-5/3}$) flanked, on the low-frequency end by a flatter branch ($\propto k^{-1}$) and, on the high-frequency end, by a much steeper branch ($\propto k^{-\alpha_1}, \alpha_1 \simeq$ 3–4). The k^{-1} branch is observed to exist approximately for the frequency range f of 4×10^{-4} to 4×10^{-3} Hz, where the spacecraft frequency $f = kV_s/2\pi$. The $k^{-5/3}$ branch is observed to exist approximately for the frequency range f of 4×10^{-3} Hz to 1 Hz. The first spectral break, therefore, lies near $f = 10^{-3}$ Hz or $k = 10^{-5}$ Km^{-1} for $V_s = 400$ Km/s and the second spectral break lies near $f = 1$ Hz or $k = 10^{-2}$ Km^{-1}. These are also the locations of the spectral breaks in all the rest of the figures. Attributing the Kolmogorov branch ($\propto k^{-5/3}$) to the standard inertial range cascade, initial explanations invoked dissipation[2] processes, in particular, the collisionless damping of Alfvén and magnetosonic waves, to explain the steeper branch ($\propto f^{-\alpha_1}, \alpha_1 \simeq$ 3–4). However, a recent critical study has concluded that damping of the linear Alfvén waves via the proton cyclotron resonance and of the magnetosonic waves by the Landau resonance, being strongly k (wave vector)-dependent, is quite incapable of producing a power-law spectral distribution of magnetic fluctuations. The damping mechanisms lead, instead, to a sharp cutoff in the power spectrum.[3] Cranmer and van Ballegooijen[4] have however, demonstrated a weaker than an exponential dependence of damping on the wave vector by including kinetic effects. However it is still steeper than that required for explaining the observed spectrum.

An alternative possibility, suggested by Ghosh *et al.*[5] links the spectral break and subsequent steepening to a "change" in the "controlling" invariants of the system in the appropriate frequency range. Stawicki *et al.*[6] have invoked the short wavelength dispersive properties of the magnetosonic/whistler waves to account for the steepened spectrum and christened it as the spectrum in the dispersion range. Krishan and Mahajan,[7,8] however, invoke the Hall effect to model the steepened part of the spectrum, and this should be correct since the steepening begins at a scale close to the ion-inertial scale, a hallmark of the Hall effect. In this chapter, the author discusses the corresponding density spectral distributions. Are the density fluctuations, produced by an independent mechanism, being convected by the velocity and the magnetic field fluctuations as a passive scalar or are they a result of the generation mechanism of other fluctuations? Some possibilities are discussed in Sec. 2. The dilemma of the density fluctuations, whether being convected by the velocity or the magnetic fluctuations at spatial scales near the ion-inertial scale, is discussed in Sec. 3, and the chapter ends with a section on conclusion.

2. Spectra of Density Fluctuations in Alfvénic Turbulence

The solar wind turbulence is modeled in terms of magnetohydrodynamic fluctuations. An assumed input spectrum k^{-1} of the Alfvén waves is believed to decay to generate the Kolmogorov $k^{-5/3}$ spectrum. Since the Alfvénic fluctuations are characterized by velocity fluctuation $\mathbf{V} = \pm \mathbf{B}$, the velocity and the magnetic fluctuations have identical spectra. The question of the origin of the k^{-1} spectrum has as yet no satisfactory answer although such a spectrum is observed under different and disparate circumstances related to self-organized criticality. One alternative is to derive the spectra using the dimensional arguments of the Kolomogorov hypotheses. The fluctuations are described in terms of nonlinear magnetohydrodynamic waves if an ambient magnetic field exists. In the absence of the ambient magnetic field one does not attempt such a description and instead an eddy description of the fluctuations is preferred. However, the two descriptions can be unified with the realization that the ambient field is nothing more than the field at the largest spatial scale, especially so when the measure of the fluctuations compares well with the measure of the ambient field as is often the case in the solar wind. The basic difference between the wave and the turbulent description of the fluctuations is in the nature of the nonlinear interactions. Whereas the near-neighbor nonlinear interactions amongst different scales or eddies have been shown to successfully furnish the Kolmogorov spectrum, the picture is far from clear for the wave–wave interaction processes. The three-wave interaction amongst the ideal Alfvénic waves is a rather restricted process without affecting the change of momentum and energy. The Hall–Alfvénic waves with their modified dispersion are a better candidate for the three-wave interaction processes. As in hydrodynamic turbulence, the spectra derived from the Kolmogorov arguments for the MHD turbulence must be ratified by the actual solutions of the MHD equations. The appearance of the k^{-1} spectrum of the Alfvén waves and the k^{-1} spectrum derived from the cascading of the magnetic helicity (as shown later) perhaps are indicators of such a ratification. The ideal MHD system supports two global invariants: the total energy E and the magnetic helicity H_M defined as

$$\text{Total Energy } E = \frac{1}{2} \int (V^2 + B^2) d^3 x = \frac{1}{2} \sum_k |V_k|^2 + |B_k|^2, \quad (1)$$

$$\text{Magnetic Helicity } H_M = \int \mathbf{A} \cdot \mathbf{B} d^3 x = \sum_k \frac{i}{k^2} (\mathbf{k} \times \mathbf{B}_k) \cdot \mathbf{B}_{-k}, \quad (2)$$

where $\mathbf{A}$ is the vector potential. In order to derive the spectral energy distributions we resort to the Kolmogorov hypotheses according to which the spectral cascades proceed at a constant rate governed by the eddy turnover time $(k\mathbf{V}_k)^{-1}$. This rate applies for MHD turbulence too if $(k\mathbf{V}_k)^{-1} \ll (kV_A)^{-1}$, where V_A is the Alfvén velocity or for strong turbulence. For ε_E denoting the constant cascading rate of the total energy E, we get the dimensional equality

$$(kV_k)V_k^2 = \varepsilon_E, \quad (3)$$

where we have made use of the Alfvénic relation $V_k = \pm B_k$. The omnidirectional spectral distribution function $W_E(k)$ of the kinetic energy per gram per unit wave vector, $(V_k^2/2k)$, then takes the form

$$W_E(k) = \frac{(\varepsilon_E)^{2/3}}{2} k^{-5/3}. \tag{4}$$

The equipartition between the kinetic and the magnetic energy yields

$$M_E(k) = W_E(k), \tag{5}$$

where $M_E(k) = (B_k^2/2k)$ is the similarly defined omnidirectional spectral distribution function of the magnetic energy density.

The cascading of the magnetic helicity H_M (ε_H being the cascading rate for the magnetic helicity) produces a different dimensional equality

$$(kV_k)\left(\frac{B_k^2}{k}\right) = \varepsilon_H, \tag{6}$$

resulting in the following different kinetic and magnetic spectral energy distributions:

$$W_H(k) = \frac{1}{2}(\varepsilon_H)^{2/3}k^{-1}, \tag{7}$$

$$M_H(k) = W_H(k). \tag{8}$$

The observed solar wind magnetic spectrum, at spatial scales much larger than the ion-inertial scale can be generated if we could string together the two branches, $M_E(k) \propto k^{-5/3}$ and $M_H(k) \propto k^{-1}$. The rationale as well as the modality for stringing different branches originates in the hypothesis of selective dissipation. It was, first, invoked in the studies of two-dimensional hydrodynamic turbulence.[9] The idea is that in a given k range, the particular invariant which suffers the strongest dissipation, controls the spectral behavior (determined, in turn, by arguments a la Kolmogorov). Thus if the k ranges associated with different invariants are distinct and separate, we have a straightforward recipe for constructing the entire k-spectrum in the extended inertial range. In 2D hydrodynamic turbulence,[9] for instance, the enstrophy invariant, because of its stronger k-dependence (and hence larger dissipation) compared to the energy invariant, dictates the large k-spectral behavior. Therefore, the entire inertial range spectrum has two segments — the energy-dominated low k and the enstrophy-dominated high $k(\propto k^{-3})$. The procedure amounts to placing the spectrum with the highest negative exponent at the highest k-end, and the one with the lowest negative exponent of k at the lowest k-end. This procedure generates the low k-end spectral branches k^{-1} and

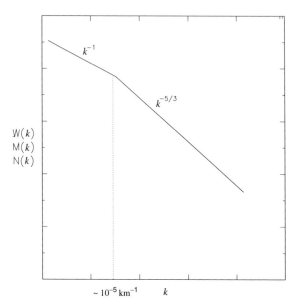

Fig. 2.   Spectral distribution of kinetic energy $W(k)$, the magnetic energy $M(k)$ and the density $N(k)$ in Alfvénic turbulence.

$k^{-5/3}$ for the kinetic energy and the magnetic energy spectrum. In Fig. 2, the break represents the energy injection scale from which the dual cascade ensues. Now, the density fluctuations if convected as a passive scalar would have the spectral distributions, $N_E(k)$ and $N_H(k)$, defined as $n_k{}^2 = kN(k)$, same as the kinetic and magnetic energy as exhibited in Fig. 2. The k^{-1} branch is observed to exist approximately for the frequency range f of 4×10^{-4} to 4×10^{-3} Hz, where the spacecraft frequency $f = kV_s/2\pi$. The $k^{-5/3}$ branch is observed to exist approximately for the frequency range f of 4×10^{-3} Hz to 1 Hz. The spectral break, therefore lies near $f = 10^{-3}$ Hz or $k = 10^{-5}$ km. However, if the density fluctuations are generated by the same mechanism as the velocity and the magnetic field fluctuations as for example in magnetosonic modes, the three spectra would be determined from their dynamical relationships. In a magnetosonic mode propagating perpendicular ($k_z = 0$) to the uniform ambient magnetic field $\mathbf{B}_0$, say in the z direction, one has

$$\|V_\perp\| = n(1 + \beta)^{1/2}, \quad \mathbf{B} = (0, 0, B_z), \quad B_z = n, \tag{9}$$

in dimensionless form. Here, n is the density fluctuation and β is the plasma β. Since the relationships amongst the three fluctuations are independent of $k_\perp$, the spectral distributions are the same as in the case of the incompressible turbulence. However, the relative magnitudes of the power in the three spectra are now given by Eq. (10). One notices that the magnetic fluctuations and the density fluctuations have identical spectra, but the velocity fluctuations can be much larger than the density or the magnetic fluctuations depending on the value of the plasma β. The plasma β in the solar wind varies from a value less than 1 to larger than 1. Thus, the relation (10) becomes useful only for large β situations. For oblique waves, the corresponding spectra would be anisotropic.

3. Spectra of Density Fluctuations in Hall-MHD

In the HALL-MHD comprising of the two-fluid model, the electron fluid equation is given by

$$m_e n_e \left[\frac{\partial \mathbf{V}_e}{\partial t} + (\mathbf{V}_e \cdot \nabla) \mathbf{V}_e \right] = -\nabla p_e - e n_e \left[\mathbf{E} + \frac{1}{c} \mathbf{V}_e \times \mathbf{B} \right]. \tag{10}$$

Assuming inertialess electrons ($m_e \leftarrow 0$), the electric field is found to be

$$\mathbf{E} = -\frac{1}{c} \mathbf{V}_e \times \mathbf{B} - \frac{1}{n_e e} \nabla p_e. \tag{11}$$

The ion fluid equation is

$$m_i n_i \left[\frac{\partial \mathbf{V}_i}{\partial t} + (\mathbf{V}_i \cdot \nabla) \mathbf{V}_i \right] = -\nabla p_i + e n_i \left[\mathbf{E} + \frac{1}{c} \mathbf{V}_i \times \mathbf{B} \right]. \tag{12}$$

Substitution for $\mathbf{E}$ from the inertialess electron equation begets

$$m_i n_i \left[\frac{\partial \mathbf{V}_i}{\partial t} + (\mathbf{V}_i \cdot \nabla) \mathbf{V}_i \right] = -\nabla (p_i + p_e) + \frac{1}{c} \mathbf{J} \times \mathbf{B}. \tag{13}$$

The magnetic induction equation becomes

$$\frac{\partial \mathbf{B}}{\partial t} = -c\nabla \times \mathbf{E} = \nabla \times (\mathbf{V}_e \times \mathbf{B}), \tag{14}$$

where $\mathbf{B}$ is seen to be frozen to electrons. Substituting for $\mathbf{V}_e = \mathbf{V}_i - \mathbf{J}/en_e$, one gets

$$\frac{\partial \mathbf{B}}{\partial t} = \nabla \times \left[\left(\mathbf{V}_i - \frac{\mathbf{J}}{en_e} \right) \times \mathbf{B} \right]. \tag{15}$$

We see that $\mathbf{B}$ is not frozen to the ions, $n_e = n_i$.

The Hall term dominates $(n_e ec)^{-1}\mathbf{J} \times \mathbf{B} \geq \mathbf{V}_i \times \mathbf{B}/c$ or the length-scale $L \leq M_A c/\omega_{pi}$ and the timescale $T \geq \omega_{ci}^{-1}$.

That is, the Hall term decouples electron and ion-motion on ion-inertial length-scales and ion cyclotron times.

The importance of the nonlinear Alfvénic state for MHD prompts one to speculate if a similar kind of an exact solution exists for Hall MHD[9,10] (HMHD), a system which encompasses MHD. We split the fields into their ambient and fluctuating parts:

$$\mathbf{B} = \hat{e}_z + \mathbf{b}; \quad \mathbf{V}_i = \mathbf{V}_0 + \mathbf{v}, \tag{16}$$

where $\mathbf{V}_0 = 0$ and substitute in Eqs. (10) and (11) to get

$$\frac{\partial \mathbf{b}}{\partial t} = \nabla \times [(\mathbf{v} - \epsilon \nabla \times \mathbf{b}) \times \hat{e}_z + (\mathbf{v} - \epsilon \nabla \times \mathbf{b}) \times \mathbf{b}], \tag{17}$$

$$\frac{\partial}{\partial t}(\nabla \times \mathbf{v}) = \nabla \times [\mathbf{v} \times (\nabla \times \mathbf{v}) + (\nabla \times \mathbf{b}) \times \hat{e}_z + (\nabla \times \mathbf{b}) \times \mathbf{b}], \tag{18}$$

where $\epsilon = \lambda_i/L$, $\lambda_i = c/\omega_{pi}$ is the ion-inertial scale with ω_{pi} the ion plasma frequency. Assuming a plane wave form $(\mathbf{V}_k, \mathbf{B}_k)\exp(ikz - i\omega t)$ we get the linear relations:

$$\mathbf{V}_k - \epsilon \nabla \times \mathbf{B}_k = -\frac{\omega}{k}\mathbf{B}_k, \tag{19}$$

$$\nabla \times \mathbf{B}_k = -\frac{\omega}{k}\nabla \times \mathbf{V}_k. \tag{20}$$

The solution of Eqs. (19) and (20) furnishes

$$\nabla \times \mathbf{B}_k = \lambda \mathbf{B}_k, \quad \lambda^2 = k^2, \quad \mathbf{B}_k = \alpha(k)\mathbf{V}_k, \quad \alpha = -\frac{\omega}{k}, \tag{21}$$

and the dispersion relation is

$$\alpha = -\frac{\epsilon k}{2} \pm \left(\frac{\epsilon^2 k^2}{4} + 1\right)^{1/2}. \tag{22}$$

One notices that the $\mathbf{V}_k$, $\mathbf{B}_k$ relation of the waves is now k-dependent, the waves are dispersive, and that they are nonlinear since for $\mathbf{b}$ given by Eq. (21), the nonlinear terms in Eqs. (17) and (18) vanish. For $k \ll 1$,

$$\alpha \to \pm 1, \quad \omega \to \mp k, \tag{23}$$

reproducing the k-independent MHD Alfvénic relationship for both the co- and the counter-propagating waves. For $k \gg 1$, it is easy to recognize, in analogy with the linear theory, that the (+) wave is the shear-cyclotron branch, while the $(-)$ represents the compressional-whistler mode. The frequency of the $(+)$ wave approaches the ion-gyro frequency. The $\mathbf{V}_k$, $\mathbf{B}_k$ relation would now give different spectral distributions for the kinetic and the magnetic energy. The Hall-MHD supports two invariants, the total energy as in the ideal MHD case and the generalized helicity, a generalized form of the magnetic helicity. The cascading characteristics of the two invariants along with the new spectral relations arising from the new $\mathbf{V}_k$, $\mathbf{B}_k$ relation are discussed in Sec. 4.

4. Generalized Helicity and Spectral Distributions in HMHD

In addition to the total energy, the HMHD system supports another invariant, the generalized helicity,[11] a generalization of the magnetic helicity H_M, defined as

$$H_G = \int d^3x (\mathbf{A} + \epsilon \mathbf{V}_i).(\mathbf{B} + \epsilon \nabla \times \mathbf{V}_i)$$

$$= \sum \left(i \frac{\mathbf{k} \times \mathbf{B}_k}{k^2} + \epsilon \mathbf{V}_k \right) \cdot (\mathbf{B}_k + i\epsilon \mathbf{k} \times \mathbf{V}_k), \tag{24}$$

where $\mathbf{A}$ is the vector potential. Notice that $H_G - H_M$ is a combination of the kinetic and the cross helicities. The generalized helicity reduces to the magnetic helicity for the vanishing Hall parameter ϵ. The cascading of the generalized helicity with a constant rate ε_G, using the relation $\mathbf{B}_k = \alpha(k)\mathbf{V}_k$ gives

$$(kV_k)[g(k)k^{-1}V_k^2] = \varepsilon_G, \tag{25}$$

$$g(k) = (\alpha + \epsilon k)^2,$$

leading to the spectral energy distributions

$$W_G(k) = \frac{1}{2}(\varepsilon_G)^{2/3}[g(k)]^{-2/3}k^{-1}, \tag{26}$$

and

$$M_G(k) = (\alpha)^2 W_G(k).$$

One notices that W_G becomes equal to W_H as H_G becomes equal to H_M for $\epsilon = 0, \alpha = 1$. The spectral distributions obtained from the cascading of the total energy are accordingly modified to

$$W_E(k) = 2^{-1/3}(\varepsilon_E)^{2/3}[1 + (\alpha)^2]^{-2/3}k^{-5/3}, \tag{27}$$

and

$$M_E(k) = (\alpha)^2 W_E(k). \tag{28}$$

It is clear that the spectral distributions (W_E, M_E, W_G, M_G) reduce to the ones obtained in the MHD case for $\epsilon = 0, \alpha = 1$. For large k, α has two roots: ϵk and $\epsilon^{-1}k^{-1}$. For $\alpha = \epsilon^{-1}k^{-1}$, corresponding to the shear Alfven wave, one can determine the kinetic and the magnetic energy spectra using again the receipe for stringing together the various spectral branches. This spectra is shown in Fig. 3 where the low k end $(k^{-1}, k^{-5/3})$ is derived in the ideal MHD regime and the high k end $(k^{-5/3}, k^{-7/3}$ for the kinetic energy and $k^{-11/3}, k^{-13/3}$ for the magnetic energy) is derived in the HMHD regime. This is in accordance with the observed magnetic spectra with the

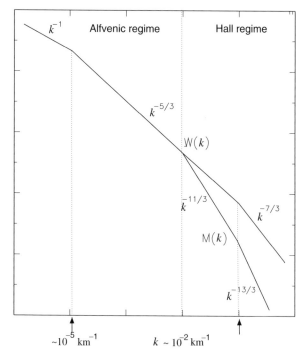

Fig. 3.   Spectral distribution of the kinetic energy $W(k)$ and the magnetic energy $M(k)$ in Hall–Alfvénic turbulence for $\epsilon = 1$.

steepened part, now, being proposed to be in the inertial range, in contrast to the other proposals, which have unsucessfully put it in the dissipation range. In the HMHD regime, the kinetic and the magnetic energy spectra are different as the two fluids now have their own dynamics. The arrows indicate the energy injection scales where the dual cascade occurs. The transition from the Alfvén to the Hall regime is smooth and does not represent a sharp break in the spectrum at the ion-inertial scale. The breaks are due to the change of invariants controlling the spectra. The spectra obtained from the second root ($\alpha = \epsilon k$) corresponding to the whistler mode would clearly be different. We find that the total energy invariant furnishes $W_E \propto k^{-3}, M_E \propto k^{-1}$, and the generalized helicity invariant furnishes $W_G \propto k^{-7/3}, M_G \propto k^{-1/3}$. Thus, the magnetic spectrum corresponding to the whistler root has no steepened branch and thus cannot account for the observed magnetic spectrum. Besides, the whistler waves undergo much stronger damping than the Alfvénic waves and may not be a major contributor to the solar wind turbulence.

5. Spectral Distribution of Density Fluctuations in HMHD

If the spectra of the kinetic and the magnetic fluctuations are different in HMHD, then which would carry the passive density fluctuations? In order to answer this question let us look at the induction equation (15) which shows that the magnetic field is frozen to the electrons. Thus, one would conclude that the electron density fluctuations would be frozen to the magnetic field fluctuations and would have the same spectrum as the magnetic field. An inspection of the induction equation in Eq. (16) shows that the magnetic field is not frozen to the ions. Thus, one would conclude that the ion density fluctuations would be carried by the velocity field fluctuations and would have the same spectrum as the velocity field. So, the electron and the ion density fluctuations have different spectral distributions without causing any violation to the quasineutrality. The entire picture is presented in Fig. 4. Summarizing, in the ideal MHD regime ($\epsilon = 0$, $\alpha = \pm 1$), the k^{-1} branch results from the cascading of the magnetic helicity, the $k^{-5/3}$ from the cascading of the total energy and the break represents the energy injection scale at which the dual cascade occurs. In the HMHD regime ($\epsilon \neq 0$, $\alpha = (\epsilon k)^{-1}$), the kinetic energy spectra $k^{-5/3}$ results from the cascading of the total energy, the $k^{-7/3}$ from the cascading of the generalized helicity and the break represents another energy injection scale at which the dual cascade occurs. All the scales lie in the inertial range of turbulence and

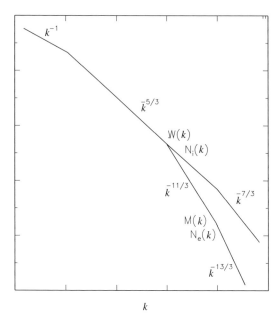

Fig. 4. Spectral distribution of the electron and the ion densities in Hall–Alfvénic turbulence.

are a consequence of the nonlinear interactions leading to the cascade. The density fluctuations are simply passive scalars carried by the velocity and the magnetic field fluctuations. The dissipation scale lies further ahead toward large k, where a more severe steepening of the spectra may exist as dictated by the damping mechanisms.

6. Conclusion

It has been shown that the observed spectral distributions of the velocity, the magnetic, and the density fluctuations in the solar wind can be modeled within the framework of the Hall magnetohydrodynamics using the dimensional arguments of the Kolmogorov hypotheses. The Hall effect is particularly needed to account for the high k-end of the spectra. The k-dependent relation between the velocity fluctuations and the magnetic field fluctuations arising from the Hall-MHD waves results in different spectra for the fluctuations. Additionally, the spectra for the electron and the ion density fluctuations, treated as passive scalars, also differ at the high k-end, again a consequence of the two-fluid treatment which brings in the Hall effect. The spectrum of the electron density fluctuations steeper than the Kolmogorov spectrum at the ion-inertial scale has been inferred from the interplanetary scintillation studies.[12] The spectrum of ion density fluctuations could be inferred from the plasma wave *in situ* measurements. Several other consequences of the foray into the two-fluid treatment such as the inclusion of compressibility and the ensuing wave modes and anisotropies need to be investigated.

Acknowledgment

Discussions with Prof. P. K. Manoharan on the density spectra are gratefully acknowledged.

References

1. M. L. Goldstein, D. A. Roberts and W. H. Matthaeus, *Ann. Rev. Astron. Astrophys.* **33** (1995) 283.
2. S. P. Gary, *J. Geophys. Res.* **104** (1999) 6759.
3. H. Li, P. Gary and O. Stawicki, *Geophys. Res. Lett.* **28** (2001) 1347.
4. S. R. Cranmer and A. A. van Ballegooijen, *Astrophys. J.* **594** (2003) 573.
5. S. Ghosh, E. Siregar, D. A. Roberts and M. L. Goldstein, *J. Geophys. Res.* **101** (1996) 2493.
6. O. Stawicki, P. S. Gary and H. Li, *J. Geophys. Res.* **106** (2001) 8273.
7. V. Krishan and S. M. Mahajan, *J. Geophys. Res.* **109** (2005) A11105.
8. V. Krishan and S. M. Mahajan, *Solar Phys.* **220** (2004) 29.
9. A. Hasegawa, *Adv. Phys.* **34** (1985) 1.
10. S. M. Mahajan and V. Krishan, *MNRAS* **359** (2005) L27.
11. Z. Yoshida and S. M. Mahajan, *Phys. Rev. Lett.* **88** (2002) 095001.
12. P. K. Manoharan, M. Kojima and H. Misawa, *J. Geophys. Res.* **99** (1994) 23411.

Advances in Geosciences
Vol. 14: Solar Terrestrial (2007)
Eds. Marc Duldig *et al.*
© World Scientific Publishing Company

ELECTRIC AND MAGNETIC FIELD VARIATIONS AT LOW AND EQUATORIAL LATITUDES DURING SC, DP 2, AND Pi 2 EVENTS

K. YUMOTO, A. IKEDA, M. SHINOHARA and T. UOZUMI

Space Environment Research Center and
Department of Earth and Planetary Sciences
Kyushu University, 6-10-1 Hakozaki, Higashi-ku
Fukuoka, 812-8581, Japan

K. NOZAKI and S. WATARI

National Institute of Information and Communication Technology
4-2-1 Nukui-Kitamachi, Koganei, Tokyo, 184-8795, Japan

K. KITAMURA

Department of Mechanic and Electrical Engineering
Tokuyama College of Technology, 3538 Kumetakajo
Shunan, 745-8585, Japan

V. V. BYCHKOV and M. SHEVTSOV

Institute of Cosmophysical Research and Radiowaves Propagation (IKIR)
Far Eastern Branch of Russian Academy of Sciences
684034 Kamchatka Region, Elizovskiy District
Paratunka, Mirnaya Str. 7, Russia

Relations of ionospheric electric and magnetic fields at low and equatorial latitudes during SC, DP 2 and Pi 2 events were investigated by analyzing the MAGDAS magnetic data and the Doppler data of our FM-CW ionospheric radar. From the analyses, we found that the ionospheric electric fields at lower latitudes during SC events consist of two components: one is the dawn-to-dusk electric field, which penetrates from the polar ionosphere into the day- and night-side equatorial ionospheres, and the other is the globally-westward electric field ($\delta \mathbf{E} = -\mathbf{v} \times \mathbf{B_0}$), which may be induced globally in the ionosphere or caused by a globally-earthward movement ($\mathbf{v}$) of the ionospheric plasmas. The H-component of DP 2 magnetic variation in the 10-nT range observed near the magnetic equator in nighttime shows roughly in-phase relation with those in daytime; however, it cannot be explained by the DP 2 dawn-to-dusk electric field observed at middle latitude in nighttime. A new generation mechanism is needed to interpret the globally coherent DP 2 magnetic variation near the magnetic equator. The compressional Pi 2 magnetic pulsation ($\delta \mathbf{B}$)

observed globally near the magnetic equator is found to show the relation of $\omega\,\delta\mathbf{B} = -\mathbf{k} \times \delta\mathbf{E}$ to the ionospheric Pi 2 electric field ($\delta\mathbf{E}$), where ω is the wave frequency, and the wave vector ($\mathbf{k}$) is directed earthward.

1. Introduction

The sudden increase of solar-wind dynamic pressure causes a sudden increase of the geomagnetic field especially at low latitudes. This phenomenon is called geomagnetic sudden commencement (SC). The disturbance field of SC is divided into two components[1]:

$$D_{SC} = DL + DP,$$

where DL is the disturbance at low latitudes and represents a step-function like increase of the H-component magnetic field at low latitudes, while DP denotes the disturbance due to polar ionospheric current, and is dominant at high latitudes. DL is caused by the current circuit flowing on the compressed magnetopause and the propagating compressional hydromagnetic (HM) wave in the magnetosphere.[2] DP shows two-pulse magnetic structure caused by the polar ionospheric electric field:

$$DP = DP_{PI} + DP_{MI},$$

where PI denotes the preliminary impulse and MI denotes the following main impulse, and the two are driven by the dusk-to-dawn and dawn-to-dusk electric fields, respectively. These electric fields are believed to penetrate from the magnetosphere into the polar ionosphere,[2,16] and instantaneously transmit into the low-latitude ionosphere by a TM (transverse magnetic) wave-guide mode.[5,6]

DP 2 magnetic fluctuations are characterized by quasi-periodic variations with timescales of about 20 min to a few hours. DP 2 magnetic fluctuations appear coherently at high latitudes and at the dayside magnetic equator. Southward turnings of the interplanetary magnetic field (IMF) are believed to be the main cause of DP 2 fluctuations on the ground.[8,9] The equivalent current system of DP 2 fluctuations consists of two vortices in the polar region, where the DP 2 dawn-to-dusk electric field penetrates instantaneously into the lower-latitude and the equatorial ionospheres.[7] At the dayside magnetic equator, ionospheric current driven by the penetrating DP 2 electric field is amplified, because of the conductivity enhancement localized within a few degrees of the dip equator (i.e., the Cowling effect). Accordingly, magnetic DP 2 fluctuations show amplitude enhancement at the dayside magnetic equator. By using this peculiarity, the dawn-to-dusk electric field in the ionosphere can be estimated from ground-based magnetic field observations. On the other hand, the ionospheric conductivity declines during nighttime. Therefore, the ionospheric current may not flow easily during the nighttime equatorial region. In this case, DP 2 electric field fluctuations in the ionosphere cannot be estimated from the magnetic observations. In order to measure the ionospheric electric fields in the nighttime, direct ionospheric observations by Frequency Modulated Continuous Wave (FM-CW) radar are needed.[22]

Pi 2 magnetic pulsations, impulsive hydromagnetic oscillations with a period of 40–150 s, occur globally at the onset of the magnetospheric substorm expansion phase. From long-term observations, it has been found that ground Pi 2 pulsations are mixtures of several components of reflecting (1) propagations of fast and shear Alfvén wave, (2) resonances of plasmaspheric/magnetospheric cavity and magnetic field lines, and (3) transformations to ionospheric current systems at low, middle, and high latitudes.[12,21] However, it has been unclear how these components coupled with each other and how their signals are distributed at different latitudes. Recently, Tokunaga *et al.*[17] explored the possibilities of identifying the global system of Pi 2 pulsations by Independent Component Analysis (ICA). They have successfully decomposed an isolated Pi 2 event on a quiet day observed at the CPMN stations into two components. One was the global oscillation that occurs from high to equatorial latitudes on the nightside with a common waveform, and has an amplitude maximum at high latitude. Another component was localized at high latitudes on the night side, whose amplitude was quite weak at low latitudes but was enhanced near the dayside magnetic equator.

The global generation and propagation mechanisms of these transient magnetic SC, DP 2, and Pi 2 variations are not yet clarified sufficiently, because they have been discussed by using only magnetic field data. Therefore, global electric and magnetic field observations and studies are needed. In order to simultaneously measure ionospheric electric fields with magnetic fields of transient SC, DP 2, and Pi 2 phenomena, and to understand how these phenomena propagate and/or penetrate globally and instantaneously from the outer magnetosphere into the lower-latitude ionosphere, the Space Environment Research Center (SERC) at Kyushu University is deploying a new MAGnetic Data Acquisition System (MAGDAS) in the Circum-pan Pacific Magnetometer Network (CPMN) region and a FM-CW radar chain at low- and mid-latitude stations along the 210° magnetic meridian (MM), as shown in Fig. 1.[23,24]

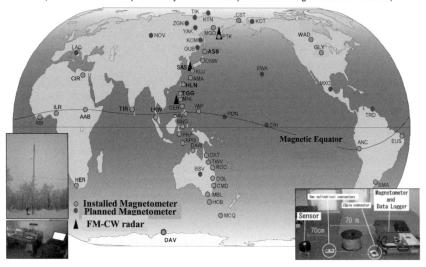

Fig. 1.　MAGDAS/CPMN station map with MAGDAS magnetometer and FM-CW radar.

2. Magnetic and Electric Field Data Set

The Circum-pan Pacific Magnetometer Network (CPMN) was constructed by Kyushu University in collaboration with about 30 international organizations along the 210° MM and the magnetic equator, respectively, during the international Solar Terrestrial Energy Program (STEP) period (1990–1997).[19–21] For space weather study and application, the SERC is now deploying a new real-time MAGnetic Data Acquisition System (MAGDAS) in the CPMN region, and a FM-CW radar array along the 210° MM. Fifty new fluxgate-type magnetometers and their data acquisition systems from Japan and overseas have been installed by the SERC since 2005.[23,24] The horizontal (H)-, declination (D)-, and vertical (Z)-components of the ambient magnetic field are measured using field-canceling coils over the dynamic range of ±64,000 nT. The magnetic variations (δH, δD, δZ) subtracted from the ambient field components (H, D, Z) are further digitized by a 16-bit A/D converter. The resolution of MAGDAS/CPMN data is 0.031 nT/LSB in the ±1000 nT range, and the estimated noise level of the magnetometers is less than 0.1 nTp-p.

In order to study SC and Pi 2 magnetic pulsations during 2002–2005, we analyzed 3-s averaged data from the CPMN stations at low latitudes, namely Kujyu (KUJ; M. Lat. 23.6°, M. Lon. 203.2°) and Kagoshima (KAG; 21.9°, 202.3°) in Japan, Pohnpei (PON; 0.08°, 229.19°) near the dip equator in the Pacific region, and Guadalupe (GLP; −0.06°, 355.57°) in Peru, and at high latitude, namely Tixie (TIK; 65.65°, 196.90°) in Russia. For the data gaps at KUJ, the data from a similar instrument at KAG were used. KUJ is about 100 km southeastward from the FM-CW radar site at Sasaguri, Fukuoka, Japan (SSG; 23.2°, 199.6°), and KAG is about 230 km southward from SSG. We also analyzed DP 2 magnetic variations observed at the MAGDAS stations at Ancon, Peru (ANC; 3.05°, 354.40°) and Santa Maria, Brazil (SMA; −19.27°, 13.29°) during daytime, and at Davao, the Philippines (−1.37°, 196.53°) during nighttime during April 2007.

In this chapter we analyzed the ionospheric data from our FM-CW radars located at Sasaguri, and Paratunka (PTK; 46.17°, 226.02°) in Kamchatka, Russia. The FM-CW radar is a type of HF (High Frequency) radar and can measure the range of target as well as its Doppler-related information. One application of the FM-CW radar is the technique developed by Barrick[3] to measure sea scatter. The FM-CW radar for the Doppler observations was first put to practical use by Poole[13] and Poole and Evans.[14] Nozaki and Kikuchi[10,11] made improvements to the FM-CW radar system. Our Doppler observations started at Sasaguri in November 2002, and at Paratunka in September 2006. From the Doppler mode observations of our FM-CW radar, we can measure vertical drift velocity and virtual height of ionospheric plasmas with high time resolution. Then, we can estimate the intensity of the ionospheric electric field. Furthermore, altitude information enables us to confirm whether the observed ionosphere is in the F-region. The sampling time of our radar system is 10 s.[4]

When the eastward electric field penetrates into the low-latitude ionosphere, the ionospheric plasma drifts upward owing to the frozen-in effect ($\mathbf{E} \times \mathbf{B}_0$ effect) in the F-region. On the other hand, the ionospheric plasma drifts downward when the westward electric field is applied into the ionosphere. Our radar system provides us with the Doppler frequency shift (Δf), that is the difference of transmitting frequency (f_0) and receiving frequency ($\Delta f + f_0$) of the radio wave reflected from the ionized layer, and a function of vertical movement of the ionized layer. The relational expression of Δf and f_0 is represented by $\Delta f = f_0 \times 2v/c$, where v is the vertical drift velocity, and c is the velocity of light. At Sasaguri we transmit radio waves of $f_0 = 8.0$ MHz during daytime and 2.5 MHz during nighttime, because the ionospheric plasma density is higher during daytime than during nighttime. From the above relational expression, the vertical drift velocity (v) of the ionosphere is given by $c\Delta f/2f_0$. The accuracy of the vertical drift velocity is 1.5 m/s and 4.7 m/s for the transmitting frequency of 8.0 MHz and 2.5 MHz, respectively.

In addition, we can calculate the intensity of the ionospheric electric fields, by using the relation of $\mathbf{E} = \mathbf{v} \times \mathbf{B_0}$, where $\mathbf{E}$ is roughly east–west electric field in the F-region, and $\mathbf{B_0}$ is the ambient magnetic field at Sasaguri. The $\mathbf{B_0}$ is given by the IGRF model (cf. http://swdcwww.kugi. kyoto-u.ac.jp/index.html at the World Data Center for Geomagnetism, Kyoto), which requires two inputs: (1) the altitude of the F-region (in this case, given by our radar), and (2) the geographical coordinates of Sasaguri (obtained from the GPS system).

3. Data Analyses of SC, DP 2, and Pi 2 Events

We have built a FM-CW radar (HF radar of 2–42 MHz) system at Sasaguri, Fukuoka. The height of dipole antenna is 26 m. HF radio wave of 2–30 MHz is emitted in the vertical direction with 20 W power in the ionosonde mode, while radio waves of central frequencies of $f_0 = 2.5$ and 8 MHz are emitted during nighttime (09–21 UT = 18–06 LT) and daytime (21–09 UT = 06–18 LT), respectively, in the Doppler mode. The speed of sweep frequency and the sampling frequency are 100–1000 kHz/s and 2000–20,000 Hz/s, respectively. This system can measure the Doppler frequency shift (Δf) of radio wave reflected from the ionized layer and the height of reflection layer at 10-s sampling rate, from which we can deduce the ionospheric electric fields associated with SC, DP 2, and Pi 2 events.

3.1. Ionospheric electric fields and ground magnetic fields during SC events

The upper panel of Fig. 2 shows SC magnetic variations at KUJ during daytime and at SMA during nighttime, and the associated ionospheric electric field observed at SSG during daytime on 4 November 2003. We can see eastward ionospheric electric field with 0.69 mV/m peak-to-peak intensity at SSG and step-like magnetic field variations of about 60 nT amplitude at both KUJ on the dayside and SMA on the nightside. The bottom panel of Fig. 2 shows the SC magnetic variation at KUJ and the ionospheric electric field at SSG observed during nighttime on

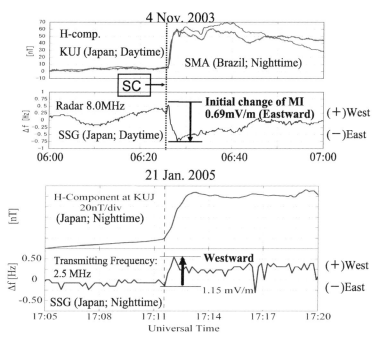

Fig. 2.   (Upper) Dayside ionospheric electric SC field at SSG with magnetic variations at KUJ during daytime and at SMA during nighttime on 4 November 2003. (Bottom) Nightside ionospheric electric SC field at SSG with magnetic variation at KUJ during nighttime on 21 January 2005.

21 January 2005. In this case, westward ionospheric electric field of 1.15 mV/m peak-to-peak intensity was observed at SSG with step-like magnetic variation of 80 nT at KUJ.

We selected 40 SC events that were identified using magnetic data from KUJ and the FM-CW radar data during the period of 2002–2005. At first, we examined step-function-like magnetic changes, and then read the peak-to-peak intensity of the ionospheric electric fields during the SC events. We found that the ionospheric electric fields denote the direction eastward during daytime (06–20 LT) and westward during the nighttime (17–07 LT), as shown in Fig. 3. The averaged peak-to-peak intensity of observed electric fields is also found to be 0.5 mV/m during the daytime and 1.0 mV/m during the nighttime. This daytime and nighttime asymmetries of observed ionospheric electric fields cannot be interpreted using only the penetration model of polar dawn-to-dusk electric field into the day- and night-side lower ionosphere during the SC events. The scale size of changes

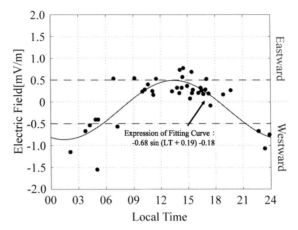

Fig. 3.   Local-time dependence of ionospheric electric fields of SCs at Sasaguri during 2002–05.

in the solar wind is too large in comparison to that of the globe; therefore, the day–night asymmetry of the ionospheric electric fields in Fig. 3 must not be related to the solar-wind conditions. We may need additional electric field component or a local time-dependence of the penetration efficiency of the polar electric fields into the low-latitude ionosphere.

We compared the peak-to-peak intensity of ionospheric electric fields with the step-function-like change of magnetic fields during SC events, and found a clear correlation (correlation coefficient = 0.70) between the electric and magnetic fields. We also compared the peak-to-peak intensity of ionospheric electric fields with the dynamic pressure (P_{sw}) in the solar wind during the interplanetary shock events. There was a correlation (correlation coefficient = 0.65) between the two, while no correlation was found between the electric field intensity in the solar wind (E_{sw}) and the intensity of ionospheric electric fields. The ionospheric electric fields seem to depend mainly on the P_{sw}.

These observations suggest that the ionospheric electric fields at low latitudes during SC events consist of two components: one is the dawn-to-dusk electric field of averaged intensity of 0.75 mV/m, which penetrates from the polar ionosphere into the day- and night-side equatorial ionospheres, and the other is the globally-westward electric field ($\delta\mathbf{E} = -\mathbf{v} \times \mathbf{B}_0$) of averaged intensity of 0.25 mV/m. The second component may be a globally induced, westward electric field ($\delta\mathbf{E}$) in the ionosphere or may be caused by a globally-earthward movement ($\mathbf{v}$) of ionospheric plasmas during SC events.

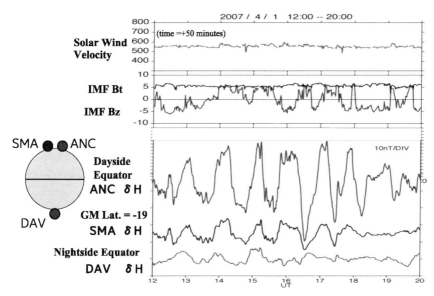

Fig. 4.   DP 2-type IMF Bz variation at ACE in the solar wind (Upper), and H-component DP 2 variations observed by MAGDAS at ANC and SMA during daytime and at DAV during nighttime (Bottom) on 1 April 2007.

3.2. *DP 2 ionospheric electric field and global magnetic field variations*

The upper panel of Fig. 4 shows the solar-wind velocity and the total and z-component of interplanetary magnetic field (IMF) observed by the ACE satellite around the Lagrange point (L1) (see http://helios.gsfc.nasa.gov/ace_spacecraft.html); the data has been shifted by 50 min forward to take into account the solar-wind velocity and the ACE satellite location. The lower panel shows the H-component magnetic field variations observed by the MAGDAS magnetometer near the magnetic equator at ANC and at low latitude at SMA during daytime and near the magnetic equator at DAV during the nighttime on 1 April 2007. Noncompressional DP 2 fluctuation with about 1 h period can be seen only in the IMF Bz component at the ACE satellite. On the other hand, the H-component DP 2 magnetic variations on the ground indicate roughly in-phase relation at lower latitudes during daytime and nighttime. If the magnetic field variations were caused by penetration electric field associated with change in IMF Bz, the penetration electric field at the nightside should have an opposite direction to that at the dayside. In particular, the three pulses between 12 and 16 UT at DAV during the nighttime show a clear in-phase relation with the dayside DP 2 variations at low and equatorial latitudes, but the nighttime DP 2 magnetic variation at 16 to 20 UT might be shielded and reduced by the sub-storm effect (not shown in Fig. 4). Also a clear equatorial enhancement of DP 2 can be seen at ANC in daytime. We do not have a definitive interpretation for the in-phase variations of DP 2 magnetic fields at the dayside and nightside magnetic equator. More observations and new models are needed to understand the observations.

Figure 5 shows the H-component DP 2 magnetic variation observed by MAGDAS near ANC during daytime, and ionospheric DP 2 electric field measured by FM-CW radar at PTK at mid-latitude during nighttime. The equatorial enhancement of dayside DP 2 magnetic variation at ANC can be driven by an ionospheric eastward electric field, while the observed ionospheric westward electric field during the nighttime is synchronized with the DP 2 magnetic variation during daytime. This observation can be interpreted by a scenario in which the IMF Bz variation drives a dawn-to-dusk electric field in the solar wind, which penetrates into the polar region and transmits into both the day- and night-sides of the low-latitude ionosphere.

It is noteworthy that the H-component DP 2 magnetic variation observed near the magnetic equator during the nighttime indicates

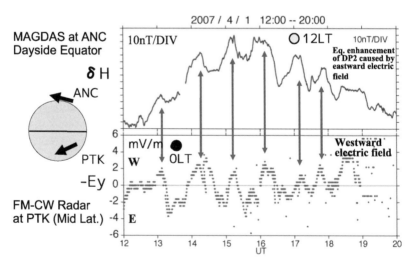

Fig. 5.   H-component magnetic DP 2 variation observed by MAGDAS at ANC during daytime, and ionospheric electric DP 2 field observed by FM-CW radar at PTK during nighttime on 1 April 2007.

roughly in-phase relation with those during daytime, as shown in Fig. 4. However, the in-phase relation cannot be explained by using the decline in conductivity during the nighttime ionosphere with the ionospheric DP 2 dawn-to-dusk electric field observed at PTK, as shown in Fig. 5. A new generation mechanism is needed to understand the globally coherent, DP 2 magnetic variations near the magnetic equator.

3.3. *Equatorial Pi 2 magnetic pulsations and ionospheric Pi 2 electric field*

The left panel of Fig. 6 shows a typical magnetic substorm event, i.e. negative and positive bay variations with Pi 2 pulsations observed at high (TIK), low (KUJ), and the equatorial (PON) latitudes on 6 November 2003. The middle panel shows the band-pass filtered amplitude–time records of H-component magnetic variations at TIK, KUJ, and PON during the nighttime and at GLP during daytime, and the ionospheric Pi 2 electric pulsation observed by the FM-CW radar at SSG in the Doppler mode during the nighttime. We can see the equatorial enhancement of Pi 2 magnetic amplitude at GLP during daytime and a globally coherent nature of Pi 2 pulsations at GLP and PON near the magnetic equator, and at

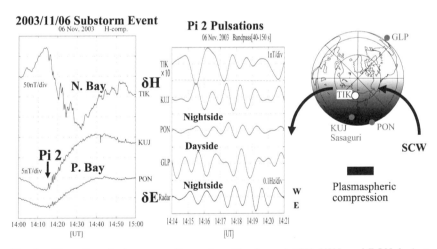

Fig. 6. Magnetic sub-storm, negative and positive bays at TIK, KUJ, and PON during nighttime (Left), Pi 2 magnetic pulsation at TIK, KUJ, PON during nighttime and at GLP during daytime, and ionospheric Pi 2 electric pulsation at SSG (Middle), and a schematic formation of sub-storm current wedge and plasma compression during the sub-storm expansion (Right).

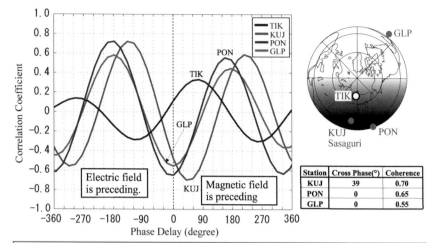

The Pi 2 electric field oscillates with the equatorial magnetic fields (PON and GLP) in-phase. Where there is a phase delay between the electric field and magnetic field at low latitude KUJ.

Fig. 7.   Cross correlation of Pi 2 magnetic pulsations at TIK, KUJ, and PON during nighttime and at GLP during daytime with ionospheric Pi 2 electric field at SSG during nighttime.

KUJ at low latitude. It is also found that the eastward electric field of ionospheric Pi 2 at SSG indicates a nearly in-phase relation with the northward magnetic field of the global lower-latitude Pi 2 pulsation.

Figure 7 shows the cross-correlation of the ionospheric Pi 2 electric field pulsation observed at low latitude (SSG) during nighttime to Pi 2 magnetic pulsation observed at high latitude (TIK), low latitude (KUJ), and magnetic equator (PON), respectively, during nighttime and at the magnetic equator (GLP) during daytime. The Pi 2 magnetic pulsations at high (TIK) and low latitudes (KUJ), respectively, show a nearly out-of-phase relation. We can see a little bit of phase difference between Pi 2 magnetic pulsations observed at the low-latitude station (KUJ) and those at the magnetic equator stations (PON, GLP). There is almost no phase delay between the eastward electric field (δE) at low-latitude station SSG during nighttime and the northward magnetic fields (δB) near the magnetic equator at GLP during daytime and at PON during nighttime. If the ionospheric conductivity is not negligible around the nighttime magnetic equator, the eastward electric field of Pi 2 may drive the ionospheric current. Further, it can produce the northward magnetic field on the ground. In the case of the conductivity decline in the nighttime ionosphere, the global compressional Pi 2 pulsation observed near the magnetic equator indicates a relation of $(\omega\,\delta\mathbf{B} = -\mathbf{k} \times \boldsymbol{\delta}\mathbf{E})$, where ω is the wave frequency and the wave vector ($\mathbf{k}$) is directed earthward.

4. Discussion and Conclusion

In order to measure ionospheric electric fields simultaneously with magnetic fields of transient SC, DP 2, and Pi 2 phenomena, and to understand how these phenomena propagate and/or penetrate globally from the outer magnetosphere into the lower-latitude ionosphere, the SERC at Kyushu University is now deploying a new MAGDAS in the CPMN region and a FM-CW radar chain along the 210°MM. From the MAGDAS/CPMN observations, the following electromagnetic characteristics of SC, DP 2, and Pi 2 events are found:

(1) The ionospheric electric fields at lower latitudes during SC events consist of two components: one is the dawn-to-dusk electric field of averaged 0.75 mV/m peak-to-peak intensity, which penetrates from the polar ionosphere into the day- and night-side equatorial ionospheres, and the other is the globally-westward electric field ($\delta\mathbf{E} = -\mathbf{v} \times \mathbf{B}_0$) of averaged 0.25 mV/m peak-to-peak intensity, which may be induced globally in the ionosphere or caused by a globally-earthward movement ($\mathbf{v}$) of the ionospheric plasmas.

(2) The H-component DP 2 magnetic variation of 10 nT observed near the magnetic equator during nighttime shows roughly an in-phase relation with those during daytime, but it cannot be explained on the basis of the ionospheric DP 2 dawn-to-dusk electric field and the conductivity decline during the nighttime ionosphere. A new generation mechanism is needed to understand the globally coherent, DP 2 magnetic variation near the magnetic equator.

(3) The global compressional Pi 2 magnetic pulsation ($\delta\mathbf{B}$) observed near the magnetic equator is found to show a relation of $(\omega\,\delta\mathbf{B} = -\mathbf{k} \times \boldsymbol{\delta}\mathbf{E})$ to the ionospheric Pi 2 electric pulsation ($\boldsymbol{\delta}\mathbf{E}$), where ω is the wave frequency and the wave vector ($\mathbf{k}$) is directed earthward.

From multi-satellite and ground magnetic field observations, Wilken *et al.*[18] proposed that the compressional HM wave-front can propagate to the nightside magnetosphere and ionosphere at the time of SC. In a recent study on SC-related electric and magnetic field perturbations in 126 SC events measured by the Akebono satellite, Shinbori et al.[15] reported that the initial electric field excursion is directed westward in the entire region of the inner magnetosphere and plasmasphere within an L-value range from 1.08 to 4.5. The intensity was in a range from 0.2 to 40 mV/m, and did not show a clear dependence on magnetic local time. Moreover, the duration of the initial electric field excursion is about 64–120 s. They also concluded that the westward electric field is an inductive field created by compressional HM wave. In the present chapter, the ionospheric electric fields observed by our FM-CW radar system at lower latitude during SC events are found to consist of two components: one is the dawn-to-dusk electric field of the Main Impulse (DP_{MI}), and the other is the globally-westward electric field ($\delta \mathbf{E} = -\mathbf{v} \times \mathbf{B}_0$), which may be induced globally in the ionosphere or caused by a globally-earthward movement ($\mathbf{v}$) of the ionospheric plasmas. These electric fields, which are induced globally in the ionosphere and/or due to globally-earthward movement of the ionospheric plasma, must be associated with the westward electric fields observed by the Akebono satellite.[15]

The H-component DP 2 magnetic variation observed near the magnetic equator during nighttime, as shown in Fig. 4, indicates roughly an in-phase relation with those during daytime. However, the observed magnetic DP 2 variations during nighttime cannot be explained by using the ionospheric DP 2 dawn-to-dusk electric field with the conductivity decline during the nighttime ionosphere. In order to understand the globally coherent, DP 2 magnetic variations near the magnetic equator, and to make a new generation model for the global DP 2 at the magnetic equator, we need more coordinated magnetic and electric field observations. For example, we can examine if the DP 2 magnetic fluctuations on the ground are driven by and/or associated with dynamic pressure changes in the solar wind and compressional variations in the magnetosphere.

Ground Pi 2 pulsations are mixtures of several components reflecting propagations of fast and shear Alfvén waves, the resonances of the magnetospheric cavity and magnetic lines of force, and transformations to the ionospheric current systems with penetrations of ionospheric electric field variations from high to equatorial latitudes.[12,21] However, there are still open questions how they couple to each other, how these signals distribute at different latitudes, and how global-mode oscillations from high to equatorial latitudes and during day- and night-time can be generated (and/or excited). We need more globally-coordinated magnetic and electric field observations and theoretical works.

Acknowledgments

The authors would like to thank Mr Ishihara, Mr Mori and Mr Shinbaru for their contributions to the construction of our FM-CW radar system. The host scientists at ANC and GLP in Peru, at SMA in Brazil, and at PON in Federated States of Micronesia are Dr T Ishitsuka, Prof Nelson J Schuch, and Prof H Utada, respectively. We also acknowledge Dr Shinbori for providing us with the SC-event list and for useful discussion. One of the authors (A Ikeda) is supported by the Professor Tatsuro Matsumoto Scholarship Fund. The PI of MAGDAS/CPMN project, K Yumoto, SERC, Kyushu University very much appreciates the 30 organizations and co-investigators around the world for their ceaseless cooperation and contribution to the MAGDAS/CPMN project. Financial supports were provided by Japan Society for the Promotion of Science (JSPS) as Grant-in-Aid for Overseas Scientific Survey (15253005, 18253005) and for publication of scientific research results (188068).

References

1. T. Araki, *Planet. Space Sci.* **25** (1977) 373–384.
2. T. Araki, in *Solar Wind Sources of Magnetospheric Ultra-Low-Frequency Waves*, eds. M. J. Engebretson *et al.*, *Geophys. Monogr. Ser.*, Vol. 81 (AGU, Washington, DC, 1994), pp. 183–200.
3. D. E. Barrick, *NOAA Technical Report* ERL 283-WPL 26 (1973).
4. A. Ikeda, K. Yumoto, M. Shinohara, K. Nozaki, A. Yoshikawa and A. Shinbori, *Mem. Fac. Sci., Kyushu Univ., Ser. D, Earth Planet. Sci.* **XXXII**(1) (2008) 1–6.
5. T. Kikuchi and T. Araki, *J. Atmos. Terr. Phys.* **41** (1979) 917–925.
6. T. Kikuchi, *J. Geophys. Res.* **91** (1986) 3101.
7. T. Kikuchi, H. Luhr, T. Kitamura, O. Saka and K. Schlegel, *J. Geophys. Res.* **101** (1996) 17161.
8. A. Nishida, *J. Geophys. Res.* **73** (1968) 1795.
9. A. Nishida, *J. Geophys. Res.* **73** (1968) 5549.
10. K. Nozaki and T. Kikuchi, *Mem. Natl. Inst. Polar Res.*, Spec. Issue **47** (1987) pp. 217–224.
11. K. Nozaki and T. Kikuchi, *Proc. NIPR Symp. Upper Atmos. Phys.* **1** (1988) 204–229.
12. J. V. Olson, *J. Geophys. Res.* **104** (1999) 17,499–17,520.
13. A. W. V. Poole, *Radio Sci.* **20** (1985) 1609.
14. A. W. V. Poole and G. P. Evans, *Radio Sci.* **20** (1985) 1617.
15. A. Shinbori, T. Ono, M. Iizima and A. Kumamoto, *Earth Planet Space* **56** (2004) 269–282.
16. T. Tamao, *Rep. Ionos. Space Res. Jpn.* **18** (1964) 16–31.

17. T. Tokunaga, H. Kohta, A. Yoshikawa, T. Uozumi and K. Yumoto, *Geophys. Res. Lett.* **34** (2007) L14106, doi: 10.1029/2007GL030174.

18. B. Wilken, C. K. Goertz, D. N. Baker, P. R. Higbie and T. A. Fritz, *J. Geophys. Res.* **87** (1982) 5901–5910.

19. K. Yumoto and the 210° MM Magnetic Observation Group, *J. Geomag. Geoelectr.* **47** (1995) 1197–1213.

20. K. Yumoto and the 210° MM Magnetic Observation Group, *J. Geomag. Geoelectr.* **48** (1996) 1297–1309.

21. K. Yumoto and the CPMN Group, *Earth Planets Space* **53** (2001) 981–992.

22. K. Yumoto, M. Shinohara, K. Nozaki, E. A. Orsco, Fr. V. Badillo, D. Bringas and the CPMN and WestPac Observation Groups, in *COSPAR Colloquia Ser.* Vol. 12, *Space Weather Study using Multipoint Techniques*, eds. Ling-Hsiao Lyu (Elsevier Science Ltd, 2002), pp. 243–247.

23. K. Yumoto and the MAGDAS Group, in *Solar Influence on the Heliosphere and Earth's Environment: Recent Progress and Prospects*, eds. N. Gopalswamy and A. Bhattachayya (2006), IBN-81-87099-40-2, pp. 399–405.

24. K. Yumoto and the MAGDAS Group, *Bull. Astron. Soc. Ind.* **35** (2007) 511–522.

Advances in Geosciences
Vol. 14: Solar Terrestrial (2007)
Eds. Marc Duldig *et al.*
© World Scientific Publishing Company

NONTYPICAL Pc5 PULSATIONS IN THE OCTOBER 2003 SUPERSTORM RECOVERY PHASE

NATALIA KLEIMENOVA and OLGA KOZYREVA

Institute of the Earths Physics, Moscow
10, B. Gruzinskaya St., Moscow, 123995, Russia

The recovery phases of the super strong magnetic storms that happened in October 2003 was accompanied by very large (up to 500–600 nT) morning and daytime Pc5 geomagnetic pulsations. The afternoon pulsations were observed with similar waveforms, intensities, and polarization in the unusually wide latitude area. The wave spectra changed with time and events but they did not change with latitudes. As a rule, the spectral maxima were latitude-independent. The main behavior of these pulsations did not coincide with that expected from the field line resonance (FLR) model which is generally accepted for the "classic" morning Pc5 wave generation. The global maps of the Pc5 amplitude distribution were computed by using the multipoint observations. The unusual deep latitude Pc5 penetration was found. The global area of the enhanced Pc5s involved the relatively narrow latitude morning band and very broad (from polar to middle/low latitudes) afternoon region. The afternoon Pc5 occurrence area was divided into high and low latitude parts, separated by the wave amplitude minimum and the wave polarization reverse. It was suggested that the considered daytime Pc5 pulsations were of non-resonant nature.

1. Introduction

It is well known that a magnetic storm recovery phase starts with the IMF Bz northward turning and the ring current decay beginning. As a rule, a storm recovery phase is accompanied by morning Pc5 geomagnetic pulsations in the frequency range of 1–6 mHz,[1–3] known as ULF (ultra-low frequency) waves. We calculate the Pc5 wave global distribution in coordinate geomagnetic latitude (LAT)–magnetic local time (MLT) by using the multi-station ground observations for the given UT time interval and the given frequency range. Figure 1 shows typical storm recovery phase Pc5 pulsations (upper panel) observed at FCC station ($\Phi = 69.8°$, $\Lambda = 329°$) and its global LAT–MLT plot (bottom panel). This map shows the distribution of pulsation intensity in given frequency band calculated

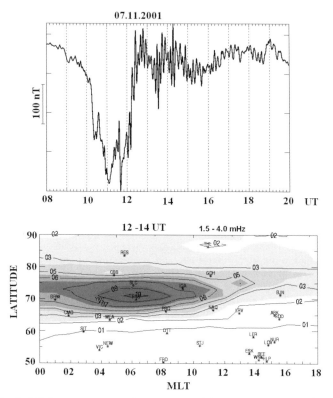

Fig. 1. Typical Pc5 geomagnetic pulsations ground-based magnetogram (upper panel) and its LAT–MLT global map distribution (bottom panel).

from the dynamic amplitude spectra in $nT/mHz^{0.5}$. One can see that the enhanced waves are observed at the relatively narrow range of the geomagnetic latitudes between $65°$ and $75°$ auroral latitudes in the dawn sector of the magnetosphere.

These pulsations are typically attributed to the field-line resonance (FLR) origin. The most likely source for FLR energy is Kelvin–Helmholtz instability (KHI) developing in the magnetosphere flanks under high solar wind velocity. Really, a good correlation between the solar wind speed and Pc5 power was found.[4,5]

The resonant frequency of a field line is determined by its length, magnetic field strength, and the local plasma density. The wave resonant frequency decreases with L-shell increasing. There should be an amplitude local peak and phase reversal across the resonant latitude.[6,7]

The FLR pulsation occurrence is asymmetric about local noon with peak occurrence at the dawn sector. Some authors[8,9] indicated the existence of multiple, discrete FLRs (waveguide/cavity modes) with remarkably stable frequencies of 1.3, 1.9, 2.6, and 3.4 mHz. In view of the dynamic nature of the magnetosphere, this extraordinary stability of FLR is difficult to be understood.

The recovery phase of the huge magnetic storm (Dst ~ -400 nT) of 29–31 October 2003 was accompanied by the Pc5 range geomagnetic pulsations with unusual properties, not matched with the FLR criteria. In this study, we present their analysis.

2. Data

The superstorm on 29–31 October 2003 (Halloween event) was caused by the extremely fast coronal mass ejection at the propagation velocity of about 2000 km/s. The Dst variations is given in Fig. 2. The recovery phase of this storm series takes place on 31 October.

2.1. *ULF geomagnetic pulsations in the storm recovery phase*

In the storm recovery phase on 31 October Pc5 pulsations were observed in the down sector as well as in the afternoon one. Some characteristics of these Pc5 pulsations were discussed in Refs. 10–12. Contrary to "classical"

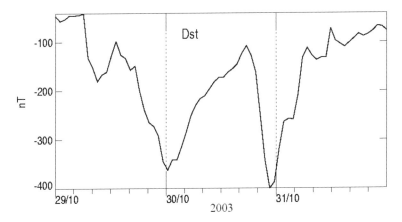

Fig. 2.   Dst-index variations on 29–31 October 2003.

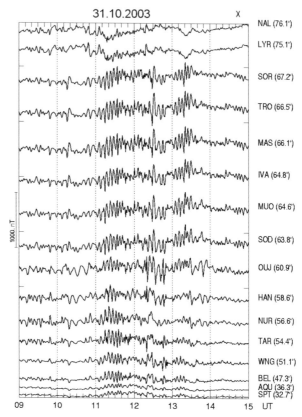

Fig. 3. Unfiltered H-component magnetograms from several stations of IMAGE meridian.

events,[5] the ULF pulsations under consideration were much stronger in the afternoon sector than in the morning one. The amplitudes of the afternoon ULF waves were as large as ∼400–500 nT under the very strong values (∼1000 km/s) of the solar wind speed.

Figure 3 demonstrates the quasi-monochromatic magnetic oscillations observed at ground-based Scandinavian IMAGE network of stations (geomagnetic noon near 09 UT) located in the afternoon sector during the late recovery storm phase on 31 October 2003. The station international cods and its geomagnetic latitudes are given on the right side.

The Pc5 pulsations with the similar waveform and intensity were recorded at very large latitude region: from SOR ($\Phi = 67°$) to SPT ($\Phi = 32°$) stations. Unfortunately, there were no observation data at the latitudes between 67° and 75°. The ULF waves did not demonstrate any phase shift with latitudes, but there was phase reversal at the latitude of ~57–58° (between HAN and NUR), at the L values of ~3.5–3.6 (probably, near the plasmapause location). Three pulsation bursts are seen in the time interval of 11–14 UT. We present below the results of the treatment of the first two bursts.

2.2. *ULF geomagnetic pulsations spectra and LAT–MLT plots*

The dynamic spectra of these ULF pulsations are presented in Fig. 4. The spectra of the different Pc5 bursts were different but their shape did not change with latitude. Even the small details were similar.

The amplitude Fourier spectra of the first two Pc5 bursts (11–12 and 12–13 UT) are presented in Figs. 5 and 6. Two main spectral maxima (at ~2.9 and ~3.6 mHz) are seen at 11–12 UT in the Scandinavian data.

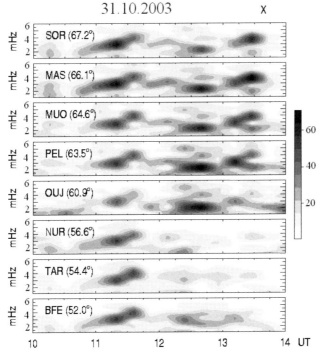

Fig. 4. The dynamic amplitude spectra of the afternoon Pc5 pulsations; the unit of the bar is $\mathrm{nT/mHz^{0.5}}$.

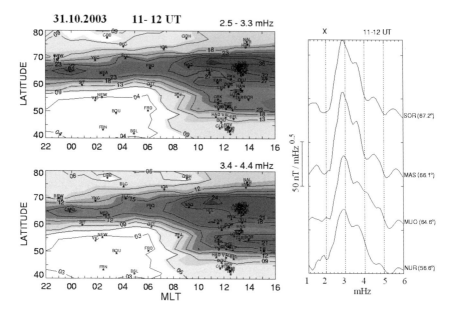

Fig. 5.   The Pc5 global LAT–MLT plots (11–12 UT) and the wave amplitude spectra.

According to dynamic spectra (Fig. 4), they did not occur simultaneously; thus, they were not the wave harmonics. We note that these picks do not fit the "magic" FRLs frequencies.[6,7] The global LAT–MLT plots (left part of Fig. 5) computed for these two frequency ranges, demonstrate a great similarity. Both band frequencies of the ULF waves have been unusually deep penetrated into the magnetosphere in the afternoon sector.

The Fourier spectra of the second ULF burst (12–13 UT) is shown in Fig. 6 (right side). Several discrete maxima are seen. The strongest pick at 2.4 mHz did not coincide with the main pick in the first ULF burst (11–12 UT). The largest amplitude of the 2.4 mHz waves was recorded at $\Phi \sim 65$–$66°$. The wave intensity decreased with latitude decreasing very rapidly. However, the wave phase did not change at the latitude of the Pc5 amplitude maximum, as it is typical for FLR. The second pick near 3.5 mHz coincided closely with the second spectral maximum of the first ULF burst (Fig. 5).

All the four maps demonstrate similar spatial distribution. In the morning sector, the Pc5 latitude distribution was typical for a storm recovery phase (see Fig. 1). However, in the afternoon the unusual spatial picture was observed. Two latitude Pc5 enhancement regions were recorded,

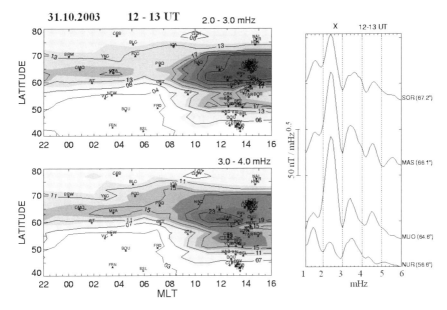

Fig. 6. The Pc5 global LAT–MLT plots (12–13 UT) and the wave amplitude spectra.

which were separated by the wave amplitude minimum and wave phase reversal between $\Phi = 57°$ and $\Phi = 58°$ (i.e. between HAN and NUR stations at 11–12 UT, Fig. 3).

2.3. *Solar wind and IMF data*

The solar wind and IMF parameters measured by ACE spacecraft during the storm recovery phase are shown in Fig. 7. At 12 UT, the estimated time delay between ACE and ground data was about 15 min. One can see that the first Pc5 pulsations burst (11–12 UT) was associated with the solar wind density jump. The second ULF burst (12–13 UT), probably, was caused by the simultaneous impulsive enhancement of the solar wind density (or dynamic pressure) and all components of IMF, observed near 12 UT.

Figure 7 shows that during the discussed interval (11–13 UT), the solar wind velocity was extremely strong (up to ~1000 km/s).

Two similar latitude areas of the afternoon Pc5 enhancement were observed in the late recovery phase of the strong magnetic storms on 29 October 2003 (Fig. 8) and 15 May 2005. It should be pointed out that

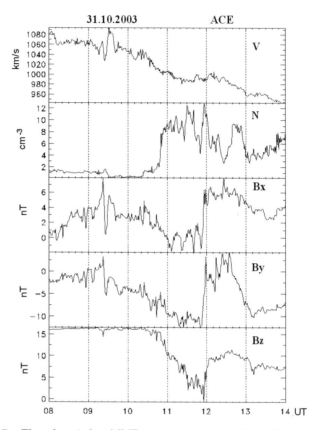

Fig. 7.   The solar wind and IMF parameters measured by ACE spacecraft.

the discussed afternoon Pc5s were observed under positive IMF Bz and a very strong solar wind velocity ($\sim$1000 km/s).

The evidence for the penetration of the afternoon ULF wave power in the Pc5 band to much lower L-values than normal was recently published by Lee et al.[13] analyzing the ULF waves during the beginning of the main phase of the very strong magnetic storm on 24 March 1991 (Dst $\sim$ −300 nT).

3. Discussion and Conclusion

The considered afternoon Pc5 range pulsations could not be associated with the "classic" (FLR) model, as it is generally accepted for the typical morning Pc5 ULF waves. Dmitriev et al.[12] interpreted the discussed ground pulsations to be due to the oscillatory motion of the magnetopause

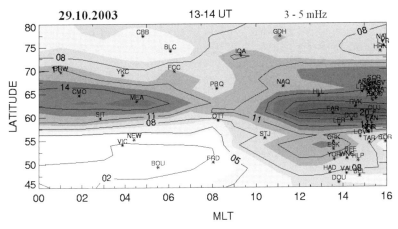

Fig. 8. The global Pc5 LAT–MLT plot in the recovery phase of the 29 October 2003 storm.

boundary layer. The following properties of the observed Pc5 pulsations do not fit the FLR model, and are inconsistent with a large number of typical Pc5 observations[2,5,7]:

(1) The Pc5 pulsations were stronger in the local afternoon than in the morning. According to the CRRES magnetic field data,[14] it is typical for the standing compressional wave (global mode) inside the magnetosphere. But such waves cannot directly reach the ground. Unfortunately, we could not compare the ground and magnetosphere data, because at the analyzed time all geostationary satellites were located at the night-side sector of the magnetosphere. Some authors,[15–17] suggested that the afternoon Pc5 pulsations observed during strong magnetic storms could be a ground signature of compressional waves in the magnetosphere. According to Fujitani *et al.*[18] the driver in this case could be solar wind dynamic pressure pulse hitting post-noon side of the magnetopause more frequently.[17,19]

(2) The Pc5 waveforms were coherent with the similar amplitudes at a very large latitude range. The spectral maxima of the wave were latitude-independent; contrary to that, the spectral maxima of FLR waves increase with the decreasing latitude.

(3) The amplitude spatial distributions of the different wave frequencies were similar. It does not fit the FLR model of wave generation.

(4) There is no phase reversal across the latitude of the Pc5 amplitude maximum as provided by the FLR theory.

(5) The Pc5 global distribution maps demonstrated the unusual deep afternoon wave penetration in the magnetosphere up to latitudes of $\Phi \sim 40$–$50°$ with two areas of the enhanced Pc5s, separated by the wave phase reversal (probably at the plasmapause?). One plausible reason for the deep wave penetration could be increased equatorial plasma mass density due to increasing heavy ions (O^+). Such heavy ion mass loading could decrease the magnetosphere eigenfrequencies. Fraser *et al.*[20] show how an increase in O^+ density can contribute to a decrease in ULF wave frequencies. According to Lee *et al.*[13] this could allow Pc5 ULF power to penetrate to low L.

4. Summary

The recovery phase of huge magnetic storms as the storm on 29–31 October 2003 (Halloween event) was accompanied by very intense afternoon Pc5 geomagnetic pulsations, which could not be attributed to the "classical" resonant waves. The unusual deep magnetosphere penetration of these Pc5 wave powers was observed. There were two latitude areas of the wave enhancement occurrences; the plasmapause might be a plausible demarcation boundary between these areas.

Thus, the Pc5 pulsations under consideration were of nonresonant nature. The solar wind density jump could provide global cavity mode exiting. But how such waves reach the ground?

Acknowledgments

The authors thank the INTERMAGNET, IMAGE, SAMNET, CARISMA, and MACCS teams for magnetometer data. This study was supported by the Program No. 16 of Russian Academy of Sciences.

References

1. V. A. Troitskaya, M. V. Melnikova, O. V. Bolshakova, D. A. Rokityanskaya and G. A. Bulatova, *Ann. USSR Acad. Sci., Earth Phys.* **6** (1965) 82.
2. L. T. Afanasyeva, *Acta Geod. Geophys. Mont. Acad Sci. Hungary* **13**(1–2) (1978) 239.
3. J. L. Posch, M. J. Engebretson, V. A. Pilipenko, W. J. Hughes, C. T. Russell and L. J. Lanzerotti, *J. Geophys. Res.* **108**(1) (2003) 1029, doi: 10.1029/2002JA009386.

4. M. J. Engebretson, K.-H. Glassmeir and M. Stellmacher, *J. Geophys. Res.* **103** (1998) 26,271.

5. G. E. Baker, E. F. Donovan and B. J. Jackel, *J. Geophys. Res.* **108**(A10) (2003) 1385, doi: 10.1029/2002JA009801.

6. L. Chan and A. Hasegawa, *J. Geophys. Res.* **79** (1974) 1024.

7. C. W. S. Ziesolleck and D. R. McDiarmid, *J. Geophys. Res.* **100**(A10) (1995) 19,299.

8. J. C. Samson, R. A. Greenwald, J. M. Ruohoniemi, T. J. Hughes and D. D. Wallis, *Can. J. Phys.* **69** (1991) 929.

9. A. D. M. Walker, J. M. Ruohoniemi, K. B. Baker, R. A. Greenwald and J. C. Samson, *J. Geophys. Res.* **97** (1992) 12,187.

10. N. G. Kleimenova and O. V. Kozyreva, *Geomagn. Aeron.* **45**(1) (2005) 75.

11. N. G. Kleimenova and O. V. Kozyreva, *Geomagn. Aeron.* **45**(5) (2005) 597.

12. A. Dmitriev, J.-K. Chao, M. Thomsen and A. Suvorova, *J. Geophys. Res.* **110** (2005) A08209, doi: 10.1029/2004JA010582.

13. E. A. Lee, I. R. Mann, T. M. Loto'aniu and Z. C. Dent, *J. Geophys. Res.* **112** (2007) A05208, doi: 10.1029/2006JA011872.

14. M. K. Hudson, R. E. Denton, M. R. Lessard, E. G. Miftakhova and R. R. Anderson, *Ann. Geophys.* **22** (2004) 289.

15. C. A. Reddy, S. Ravindran, K. S. Viswanathan, B. V. K. Murthy, D. R. K. Rao and T. Araki, *Ann. Geophys.* **12** (1994) 565.

16. J. J. Schott, N. G. Kleimenova, J. Bitterly and O. V. Kozyreva, *Earth, Planet Space* **50** (1998) 101.

17. G. Chisham and D. Orr, *J. Geophys. Res.* **102**(A11) (1997) 24,339.

18. S. T. Fujitani, T. Araki, K. Yumoto, S. Shiokawa, S. Tsunomura and Y. Yamada, *STEP GBRESC News* **3** (1993) 15.

19. R. A. Mathie, I. R. Mann, F. W. Menk and D. Orr, *J. Geophys. Res.* **104** (1999) 7025.

20. B. J. Fraser, J. L. Horwitz, J. A. Slavin, Z. C. Dent and I. R. Mann, *Geophys. Res. Lett.* **32** (2005) L04102, doi: 10.1029/2004GL021315.

Advances in Geosciences
Vol. 14: Solar Terrestrial (2007)
Eds. Marc Duldig *et al.*
© World Scientific Publishing Company

POLARIZATION PROPERTIES OF THE ULTRA-LOW FREQUENCY WAVES IN NON-AXISYMMETRIC BACKGROUND MAGNETIC FIELDS

K. KABIN*, R. RANKIN, I. R. MANN and A. W. DEGELING

Department of Physics, University of Alberta
Edmonton, AB, Canada T6G 2G7
*kabin@phys.ualberta.ca

S. R. ELKINGTON

LASP, University of Colorado
Boulder, CO 80303, USA

In this chapter, we present results concerning periods and polarizations of the fundamental modes, and their first two harmonics of cold plasma ultra-low frequency (ULF) guided Alfvén waves in a non-axisymmetric geomagnetic field. The background magnetospheric magnetic field is approximated by a compressed dipole which is describable in terms of Euler potentials. In this study, we focus on the polarization properties of the ULF waves because they are very important for energization of outer-radiation belt electrons, which is the ultimate motivation for the present work.

1. Introduction

Geomagnetic pulzations in the Pc-5 range are commonly observed in ground-based magnetometer data and often can be interpreted as standing shear Alfvén waves.[1] Such ultra-low frequency (ULF) waves provide an important energization mechanism for relativistic electrons in the outer radiation belt.[2–6] However, electron energization may occur only if the wave electric field has a component along the trajectory of the particle drift.[7] Thus, polarization of the ULF waves is extremely important for the effectiveness of the energization process. Energization of electrons in the MeV energy range by ULF waves was considered by Elkington et al.[4] who assumed that Alfvén waves had either purely radial or purely azimuthal polarization. Both modes contributed to electron energization in the non-axisymmetric background field of Ref. 4 because the unperturbed electron trajectories (the lines of constant magnetic field magnitude) were not circular.

A recent model of shear Alfvén waves in non-axisymmetric fields[8] allows a self-consistent calculation of both frequencies and polarizations of the waves in cold plasma. This model has been applied to a three-dimensional extension of the two-dimensional compressed dipole field of Ref. 4 to study the properties of the fundamental modes of Alfvénic waves.[9] In the present work, we continue to use the same background magnetic field introduced by Kabin et al.[9] and extend our analysis to include not only the fundamental mode, but also the first and second harmonics. We find that the polarization properties of the harmonics are different from those of the fundamental modes and they strongly depend on the magnetic local time (MLT).

2. Model Description

Following Ref. 9, we describe the background magnetic field $\mathbf{B}$ in terms of Euler potentials α and β as follows:

$$\mathbf{B} = \nabla\alpha \times \nabla\beta;$$

$$\alpha = \frac{B_0}{r}\sin^2\theta - \frac{1}{2}r^2 b_1(1 + b_2\cos\phi)\sin^2\theta, \quad \beta = \phi, \tag{1}$$

$$\mathbf{B} = \mathbf{e}_r\left(\frac{2B_0}{r^3} - b_1(1 + b_2\cos\phi)\right)\cos\theta$$

$$+ \mathbf{e}_\theta\left(\frac{B_0}{r^3} + b_1(1 + b_2\cos\phi)\right)\sin\theta. \tag{2}$$

Here r, θ, ϕ are the usual spherical coordinates with r measured in Earth radii and B_0 is the equatorial strength of the dipole field at $r = 1$. Parameters b_1 and b_2 have the following physical meaning: b_1 is related to the IMF strength and b_2 is related to the solar wind dynamic pressure. Obviously, for small r the magnetic field given by (1) asymptotically approaches that of a dipole field. The same limit is achieved as $b_1 \to 0$. Although very simple, the compressed dipole topology provides a qualitative approximation to the real magnetospheric field and has many realistic features. For example, the magnetic field is compressed on the days-side and stretched on the night-side, see figures in Ref. 9. This field is clearly not axisymmetric. At the equator ($\theta = \pi/2$) field (2) becomes $B(r, \phi) = B_0/r^3 + b_1(1 + b_2\cos\phi)$ which is the same as the compressed dipole of Refs. 4 and 10.

Alfvén wave studies also require the plasma mass density distribution along the field lines. In this study, we approximate the plasma density by a simple power law $\rho = \rho_{eq}(r/5)^{-4}$ throughout the magnetosphere, similarly to many previous studies.[9,11,12] Here r is the distance from the center of the Earth, measured in Earth radii, and $\rho_{eq} = 7\,\text{amu/cm}^3$, is the equatorial plasma density at $L = 5$. Although different density profiles were occasionally used in the past,[13,14] the details of the density distribution have relatively small effects on the properties of standing shear Alfvén waves[15] and are unimportant for the present study, which focuses on qualitative properties of the ULF waves.

Once the magnetic field and plasma density are known, the frequency and polarization of the standing shear Alfvén waves can be calculated for any particular field line from the following set of equations[8]:

$$\frac{1}{\sqrt{g}}\frac{\partial \delta B_2}{\partial \mu} = \frac{1}{v_A^2}\left(g^{11}\omega\delta E_1 + g^{12}\omega\delta E_2\right), \tag{3}$$

$$\frac{1}{\sqrt{g}}\frac{\partial \delta B_1}{\partial \mu} = -\frac{1}{v_A^2}\left(g^{21}\omega\delta E_1 + g^{22}\omega\delta E_2\right), \tag{4}$$

$$\frac{1}{\sqrt{g}}\frac{\partial \delta E_1}{\partial \mu} = -\left(g^{12}\omega\delta B_1 + g^{22}\omega\delta B_2\right), \tag{5}$$

$$\frac{1}{\sqrt{g}}\frac{\partial \delta E_2}{\partial \mu} = \left(g^{11}\omega\delta B_1 + g^{12}\omega\delta B_2\right). \tag{6}$$

Here g^{11}, g^{12}, and g^{22} are components of the metric tensor and g is the determinant of the inverse tensor for the field-aligned coordinate system. The main difficulty of the analysis lies in defining a suitable field-aligned coordinate system, which in general has to be non-orthogonal.[16] For compressed dipole (2), two coordinates of such a coordinate system are conveniently provided by the Euler potentials (1), which remain constant along any particular field line. The third coordinate, μ, which varies along the field line may be chosen as $\mu = \cos\theta/r^2$. A coordinate system defined in this manner has an advantage of becoming orthogonal for the pure dipole field, when it coincides with the usual dipole coordinates.[1,8] The two-parameter eigenvalue problem for the system of Eqs. (3)–(6) which allow us to compute both the frequency and polarization of the standing shear Alfvén wave is described in Ref. 8. In the uncoupled axisymmetric limit, these equations describe two decoupled solutions for the toroidal (zero azimuthal wave number) and poloidal (infinite azimuthal wave number)

guided Alfvén modes.[9] Note, that the often-used model for shear Alfvén
waves in a general magnetic field topology[17] is not self-consistent as the
polarization of the wave has to be artificially imposed.[16,18] In general, metric
coefficients g^{11}, g^{12}, and g^{22} have to be computed numerically,[8,18] however,
for compressed dipole field we have explicit analytical expressions for
the Euler potentials which significantly simplify the numerical procedure
and allow us to minimize numerical errors.[9] Note that our ULF wave
model does not include any excitation mechanism; it only describes the
periods and polarizations of the Alfvénic oscillations of individual field
lines assuming that there is no coupling between them. Our model also
does not provide information about the amplitudes of the standing shear
Alfvén waves nor how the wave fields change with radius and azimuthal
angle. The eigensolutions considered in this work are generalizations of the
well-known guided toroidal and poloidal modes which exist in dipole field
and are generally interpreted as representing time-asymptotic states of the
Alfvénic oscillations in the magnetosphere.

3. Results

This study focuses on the properties of the higher harmonics of standing
shear Alfvén waves in compressed dipole geometry. Figure 1 shows the
periods of the fundamental modes and their two harmonics in the
compressed dipole background field for $B_0 = 31000\,\text{nT}$, $b_1 = 10\,\text{nT}$,
and $b_2 = 8$. These parameters were chosen for consistency with the
previous study.[9] Changing the values of parameters b_1 and b_2 leads only to
quantitative modifications of the ULF properties. For example, as either
b_1 or b_2 are reduced, the background field becomes more dipolar, and
the periods of the guided Alfvén waves become more axisymmetric in the
equatorial plane. The two upper panels of Fig. 1 show the periods of the
two fundamental modes; the left one corresponds to the mode with poloidal
electric field at midnight meridian, and the right one to the mode with
toroidal electric field at midnight. Harmonics of these modes are shown in
the corresponding columns of Fig. 1: the two middle panels correspond to
the first harmonic, and the lower two panels to the second harmonic. White
points on the four upper panels show the location of the point where the
periods of the two modes coincide. For the second harmonic, instead of a
single point, there is a contour (white line) where the periods of the two
modes are the same. Similarly to Ref. 9 the lines of constant period in Fig. 1
for either the fundamental modes or their harmonics do not coincide with

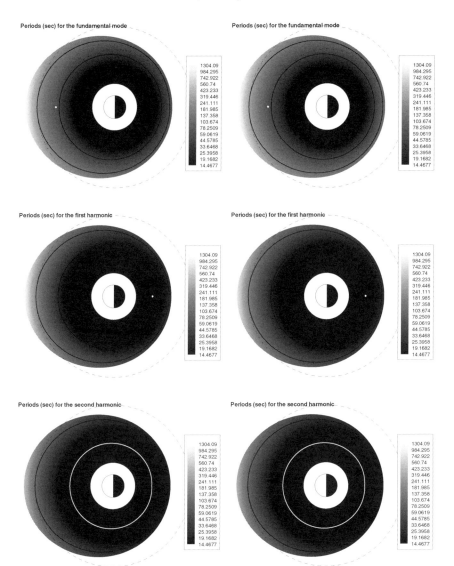

Fig. 1.   Periods of the different Alfvénic modes and their harmonics in compressed dipole field with $B_0 = 31,000\,\text{nT}$, $b_1 = 10\,\text{nT}$, and $b_2 = 8$. Dashed lines are circles with $r = 2$, 4, and 6, respectively. Bold solid lines are the contours of constant B initiated at $r = 2$, 4, and 6 at noon. White dots and contours show the locations where the periods of the two modes coincide.

the lines of constant magnetic field (shown by bold solid lines in Fig. 1). This fact reflects important modifications required for some of the original assumptions of Ref. 4. Figure 1 also shows that for higher harmonics of different polarizations the differences between the periods are much smaller than for the fundamental modes. This is not surprising since as the order of the harmonic increases, its period is described more accurately by the WKB approximation. This approximation is obviously the same for both modes and does not depend on the polarization of the wave. This fact, unfortunately, complicates the numerical calculation of the periods, since finding two almost coincident roots of an equation is a more challenging problem than finding two clearly distinct roots. For this reason, in the present work we limit our study to the first two harmonics.

The distribution of electric and magnetic fields along the field line for the first three harmonics is shown in Fig. 2. As an example we used a field line at midnight at $L = 7$ and the same parameters for the compressed dipole field as before. At midnight and at noon the off-diagonal coefficients in Eqs. (3)–(6) disappear because of the local symmetry and the two modes become uncoupled, as described in Refs. 8 and 9. Thus, in the symmetry plane of the compressed dipole field only the modes with either purely toroidal or purely poloidal polarizations are possible. Figure 2 shows modes with poloidal electric and toroidal magnetic fields; modes with poloidal magnetic and toroidal electric fields look very similar to these. Note that, at the equator, the fundamental mode and its second harmonic have a node for the magnetic field, and an antinode for the electric field. Therefore, we use equatorial electric field to characterize the polarizations of these harmonics. In contrast, the first harmonic has an electric field node and magnetic field antinode in the equatorial plane. Thus, we use the equatorial magnetic field to refer to the polarization of this harmonic.

Figure 3 shows the polarizations in the equatorial plane of the fundamental modes and their two harmonics in the compressed dipole background field. The upper panels show the fundamental modes (polarization based on the electric field), the middle panels show the first harmonic (magnetic field), and the lower panels show the second harmonic (electric field polarization). In estimating the effectiveness of the waves for electron energization outside the equatorial plane it is important to remember that the direction of the wave electric field is roughly orthogonal to the direction of the wave magnetic field. In other words, a wave with a mostly azimuthal electric field will have a mostly meridional magnetic field, and vice versa. In axisymmetric background magnetic fields ULF waves are

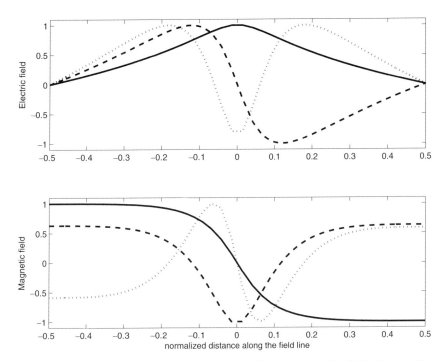

Fig. 2. Electric and magnetic fields wave distribution along the field line for the fundamental mode and its first two harmonics. The field line is at midnight, $L = 7$. Compressed dipole parameters: $B_0 = 31{,}000\,\text{nT}$, $b_1 = 10\,\text{nT}$, and $b_2 = 8$. Solid line: fundamental mode; dashed line first harmonic; dotted line: second harmonic.

usually referred to as either poloidal or toroidal (a classification which is for historical reasons based on the magnetic field perturbation), however this classification is generally not appropriate for non-axisymmetric fields.[9,18]

The polarization properties of the harmonics change along the contours of constant magnetic field intensity in a manner qualitatively similar to the fundamental mode. Along the noon and midnight meridians, only modes with either purely meridional or purely azimuthal polarizations are possible[9] for either the fundamental modes or their harmonics. It is interesting to note, however, that for some contours of constant magnetic field the polarization of the second harmonics changes twice (for example, near points marked 1 and 2 in the lower right panel of Fig. 3): a mode may start with an azimuthal electric field at midnight, then change to a mostly radial electric field at the dawn or dusk, and then become purely radial again at noon. This is a new type of polarization behavior, which has

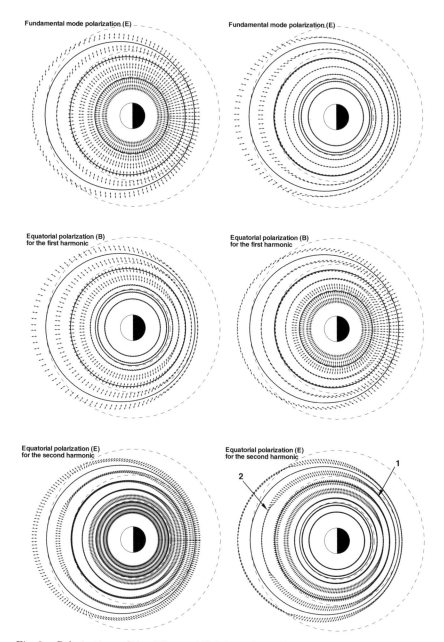

Fig. 3.   Polarizations of the different Alfvénic modes and their harmonics in compressed dipole field with $B_0 = 31{,}000\,\mathrm{nT}$, $b_1 = 10\,\mathrm{nT}$, and $b_2 = 8$. Dashed lines are circles with $r = 2$, 4, and 6, respectively. Bold solid lines are the contours of constant B initiated at $r = 2$, 4, and 6 at noon.

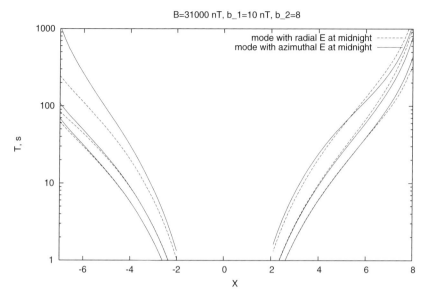

Fig. 4. Periods of the fundamental modes and their two harmonics along the noon–midnight line. Positive X is toward the Sun.

not been previously reported in the studies of the fundamental modes. As discussed in Ref. 9 and seen in Fig. 4, for the fundamental mode there is a point r_* on the noon meridian, beyond which mode with radial E has a larger period than the mode with azimuthal E. For the selected values of the parameters B_0, b_1, and b_2, $r_* = 5.285$ (white dot in the upper two panels of Fig. 1). For $r < r_*$ at noon, and for all radial distances at midnight, the period of the mode with radial E is larger than that for the mode with azimuthal E. As discussed in Ref. 9 this means that the polarization of the mode has to change from radial to azimuthal (or vice-versa), as one moves along the contour of constant period from noon to midnight for $r > r_*$. The situation is similar for the first harmonic, except that the critical point where the periods of the two modes coincide is located at midnight (white dot in the two middle panes of Fig. 1). For the second harmonic, however, instead of an isolated point, there is a contour in the equatorial plane where the periods of the two modes coincide, shown in white in the lower panels of Fig. 1. Therefore, for the second harmonic, the polarizations at noon and midnight are the same for either mode, but can vary significantly for other local times. This explains the double change of the polarization seen at certain distances for the second harmonic seen in Fig. 3.

4. Conclusions

In this chapter we extended an earlier study to include the first two harmonics in addition to the fundamental modes of the guided Alfvén waves in non-axisymmetric magnetic fields. We find that for the most part the behavior of the polarizations of the ULF wave harmonics in the compressed dipole magnetic field is qualitatively similar to that of the fundamental modes. The polarization of the harmonics changes considerably along the lines of constant background magnetic field, which has important consequences for the drift resonance mechanism of acceleration of radiation belt electrons. Electrons drifting along a line of constant magnetic field intensity in non-axisymmetric magnetosphere will experience only localized areas of accelerating electric fields. Compared to the wave fields produced by the fundamental modes, the electric fields of the higher harmonics appear to be even more fragmented and localized in space. Furthermore, because the periods of the waves (both the fundamental modes, and their harmonics) change along the unperturbed drift trajectory of electrons, the amplitude of the energizing electric field will vary considerably along the lines of constant magnetic field.

In the next stage of our work, we will develop a complete model of the ULF waves in a non-axisymmetric magnetosphere. This model will be similar to Refs. 19 and 20 and use an expansion in terms of the eigenfunctions (which are analyzed in the present work) along the field lines and couple these eigenfunctions at different locations with compressional Alfvén waves. This model will also include an external driver used as an excitation mechanism, and would be suitable for studies of specific events. It will also be used to study electron energization directly using self-consistent physics of the ULF waves.

Acknowledgment

This work was supported by the Canadian Space Agency and by Natural Sciences and Engineering Research Council of Canada, including NSERC Discovery Grant awarded to J.C. Samson.

References

1. J. Samson, Geomagnetic pulsations and plasma waves in the Earth's magnetosphere, in *Geomagnetism*, Vol. 4, ed. J. A. Jackobs (Academic Press, New York, 1991), pp. 481–592.

2. G. Rostoker, S. Skone and D. N. Baker, *Geophys. Res. Lett.* **25** (1998) 3701.
3. R. A. Mathie and I. R. Mann, *Geophys. Res. Lett.* **27** (2000) 3261.
4. S. R. Elkington, M. K. Hudson and A. A. Chan, *J. Geophys. Res.* **108** (2003), doi: 10.1029/2001JA009202.
5. T. O'Brien, K. R. Lorentzen, I. R. Mann, N. P. Meredith, J. B. Blake, J. F. Fennell, M. D. Looper, D. K. Milling and R. R. Anderson, *J. Geophys. Res.* **108** (2003), doi: 10.1029/2002JA009784.
6. S. R. Elkington, A review of ULF interactions with radiation belt electrons, in *Magnetospheric ULF Waves: Synthesis and New Directions*, eds. K. Takahashi, P. J. Chi, R. E. Denton and R. L. Lysak (AGU, 2006), pp. 177–193.
7. T. G. Northrop, *The Adiabatic Motion of Charged Particles* (Interscience Publishers, New York, 1963).
8. R. Rankin, K. Kabin and R. Marchand, *Adv. Space Res.* **38** (2006) 1720.
9. K. Kabin, R. Rankin, I. R. Mann, A. W. Degeling and R. Marchand, *Ann. Geophys.* **25** (2007) 815.
10. S. R. Elkington, M. K. Hudson and A. A. Chan, *Geophys. Res. Lett.* **26** (1999) 3273.
11. C. L. Waters, J. C. Samson and E. F. Donovan, *J. Geophys. Res.* **101** (1996) 24,737.
12. Z. C. Dent, I. R. Mann, F. W. Menk, J. Goldstein, C. R. Wilford, M. A. Clilverd and L. G. Ozeke, *Geophys. Res. Lett.* **30** (2003), doi: 10.1029/ 2003GL016946.
13. D.-H. Lee and R. L. Lysak, *J. Geophys. Res.* **94** (1989) 17,097.
14. R. E. Denton, J. Goldstein, J. D. Menietti and S. L. Young, *J. Geophys. Res.* **107** (2002), doi: 10.1029/2001JA009136.
15. D. Orr and J. A. D. Matthew, *Planet. Space Sci.* **19** (1971) 897.
16. A. Salat and J. A. Tataronis, *J. Geophys. Res.* **105** (2000) 13,055.
17. H. J. Singer, D. J. Southwood, R. J. Walker and M. G. Kivelson, *J. Geophys. Res.* **86** (1981) 4589.
18. K. Kabin, R. Rankin, C. L. Waters, R. Marchand, E. F. Donovan and J. C. Samson, *Planet. Space Sci.* **55** (2007) 820.
19. A. W. Degeling, R. Rankin, K. Kabin, R. Marchand and I. Mann, *Planet. Space Sci.* **55** (2007) 731.
20. A. W. Degeling, L. Ozeke, R. Rankin, I. R. Mann and K. Kabin, *J. Geophys. Res.* **113** (2008), doi: 10.1029/2007JA012411.

Advances in Geosciences
Vol. 14: Solar Terrestrial (2007)
Eds. Marc Duldig *et al.*
© World Scientific Publishing Company

S–M–I COUPLING DURING THE SUPER STORM
ON 20–21 NOVEMBER 2003

T. NAGATSUMA and K. T. ASAI

*Space Environment Group, National Institute of Information
and Communications Technology, 4-2-1 Nukui-kita
Koganei, Tokyo 184-8795, Japan*

R. KATAOKA

*Computational Astrophysics Laboratory
RIKEN, 2-1 Hirosawa, Wako, Saitama, 351-0198, Japan*

T. HORI and Y. MIYOSHI

*Solar-Terrestrial Environment Laboratory
Nagoya University, Furo-cho, Chikusa-ku
Nagoya, Aichi 464-8601, Japan*

We investigate how the solar wind, the magnetosphere, and the ionosphere
(S–M–I) coupling drives magnetospheric convection and the geomagnetic
storm during a super storm event on 20–21 November 2003. The cross-
polar cap potential (CPCP) estimated from the solar wind parameters based
on Siscoe–Hill model is consistent with those estimated from the northern
polar cap (PCN) index. It is clarified that the magnetospheric convection
is highly saturated during this event. On the contrary, the variations of the
geomagnetic storm represented by the Dst index develops linearly even during
this extreme event. Injection rate of the solar wind energy (Q) estimated
from the solar wind parameters corresponds with those estimated from Dst.
However, during the first half of the storm main phase, both CPCP and Q
estimated from geomagnetic indices is less than those estimated from the solar
wind parameters. The intensity of the magnetic field in the magnetosheath is
small relative to those in the second half of the storm main phase because of the
decreasing of Alfven Mach number below 2 within the intense magnetic cloud.
We conclude that the S–M–I coupling drives the magnetospheric convection
and the geomagnetic storm variation in different manners even in the extreme
conditions, and the efficiency of S–M–I coupling decreases when the Alfven
Mach number is less than 2.

1. Introduction

The coupling among the solar wind, the magnetosphere, and the ionosphere
(S–M–I coupling) is important for space weather research. Magnetospheric

convection and geomagnetic storm are typical manifestations of S–M–I coupling. However, the S–M–I coupling on each phenomenon seems to behave differently. The development of geomagnetic storm (ring current) shows almost linear development, as suggested previously.[1] On the contrary, the development of cross-polar cap potential (CPCP), which is the manifestation of the magnetospheric convection, shows saturation.[2] Studying the S–M–I coupling on each phenomenon during the extreme conditions of the solar wind is important, since most of the previous studies have been done by the usual conditions of the solar wind.

To examine the linearity/nonlinearity and the efficiency of the S–M–I coupling during the super storm event on 20–21 November 2003, which was occurred under the extreme conditions of the solar wind, we estimate CPCP and Q from three and two different methods, respectively. We show the efficiency of the S–M–I coupling on each phenomenon is decreasing when the Alfven Mach number is less than 2.

2. Super Storm Event on 20–21 November 2003

A super storm that occurred on 20–21 November 2003 is the largest storm in the solar cycle 23, and is the fifth largest storm (min.Dst: $-422\,\mathrm{nT}$) in terms of the Dst index observed between 1957 and 2006. The solar wind cause was a magnetic cloud with the largest southward interplanetary magnetic field observed during solar cycle 23.[3] The solar wind electric field is also the largest during this event. This is the good opportunity to study the S–M–I coupling during the extreme conditions of the solar wind with many kinds of data from satellite and ground. We introduce several methods to estimate CPCP and Q for examining the extreme event study, using the solar wind parameters and geomagnetic indices.

2.1. Estimation of cross-polar cap potential

Cross-polar cap potential (CPCP) can be estimated from several methods using ground and/or satellite observations. In this subsection, we introduce three methods for estimating CPCP.

(1) CPCP from solar wind electric field with a fixed geoeffective scale size

CPCP is estimated from the solar wind electric field and fixed geoeffective scale length using an equation as follows:

$$\Phi_{PC} = L_G \times E_m, \tag{1}$$

where L_G is a geoeffective scale length. For the case of Burke *et al.*,[4] L_G is 4.5R_E. E_m is the merging electric field, which was proposed by Kan and Lee.[5] Based on this equation, the solar wind electric field and CPCP have a linear relationship, when the geoeffective scale length is fixed.

(2) *CPCP from Siscoe–Hill model with the solar wind parameter and F10.7*

Siscoe *et al.*[6] introduce the following equations to explain the saturation of CPCP;

$$\Phi_{PC} = 57.6 E_m P_{SW}^{1/3} \left/ \left(P_{SW}^{1/2} + 0.0125 \xi \Sigma_P E_m \right), \right. \tag{2}$$

$$\xi = 4.45 - 1.08 \log \Sigma_P, \tag{3}$$

where P_{SW} is the solar wind dynamic pressure. ξ is a coefficient that depends on the geometry of the currents flowing in the ionosphere. Σ_P is the Pedersen conductivity. For this study, Pedersen conductivity is estimated from the following equation used by Nagatsuma[7];

$$\Sigma_P = 1.2 S_a^{0.5}((\cos \chi_N + \cos \chi_S) + 0.3472) + 5.0, \tag{4}$$

where χ_N and χ_S are the solar zenith angles for northern and southern CGM pole, respectively. S_a is the F10.7 index.

(3) *CPCP from a geomagnetic activity index*

The northern polar cap (PCN) index has a good correlation with CPCP. Troshichev *et al.*[8] derived the following equation:

$$\Phi_{PC} = 19.35 \times \text{PCN} + 8.78. \tag{5}$$

Based on Eq. (5), CPCP is estimated only from PCN, a geomagnetic activity index.

2.2. *Estimation of the injection rate for the ring current development*

The injection rate (Q) for the ring current development can be estimated from several methods using ground and/or satellite observations. In this subsection, we introduce two methods for estimating CPCP.

(1) Q *from the solar wind electric field*

Using the solar wind parameter, Q is estimated from the following equation introduced by O'Brien and McPherron[1]:

$$Q = -4.4(VB_S - 0.49), \qquad (6)$$

where V is the solar wind velocity, B_S is the southward component of interplanetary magnetic field. Equation (6) suggests a linear relationship between the solar wind electric field and Q. We can estimate Q from the solar wind data using this equation.

(2) Q *from a geomagnetic index*

Using the Dst index, Q is estimated from the following equations introduced by O'Brien and McPherron[1]:

$$Q = d\mathrm{Dst}/dt + \mathrm{Dst}/\tau, \qquad (7)$$

$$\tau = 2.4 \exp[9.74/(4.69 + VB_S)], \qquad (8)$$

where τ is the time constant for the ring current decay. Using these equations, we can estimate Q from a geomagnetic index.

2.3. *Results and discussion*

Figure 1 shows the S–M–I coupling during the super storm event on 20–21 November 2003. In this figure, OMNI-2 data is used as solar wind data.[9] Top panel shows the variations of CPCP estimated from the three different methods. The middle panel shows the variations of injection rate estimated from the two different methods. The bottom panel shows the Dst index.

CPCP estimated from PCN is significantly weaker than those estimated from the solar wind electric field with fixed scale length during the main phase of the storm. One of the possibilities for the significant weakening is that the geoeffective scale size during the storm time is smaller than that we assumed in our estimation. The size of the magnetosphere is decreasing due to the enhancement of the dynamic pressure during storm time. However, this effect can explain only about a half of decreasing of CPCP. It is suggested that major part of the difference is caused by the saturation of CPCP under the extreme conditions of the solar wind electric field. CPCP estimated from PCN and that estimated from Siscoe–Hill model corresponds well in general. However, the difference can be seen in the first half of the main phase. CPCP from Siscoe–Hill model is larger than those from PCN during this time period.

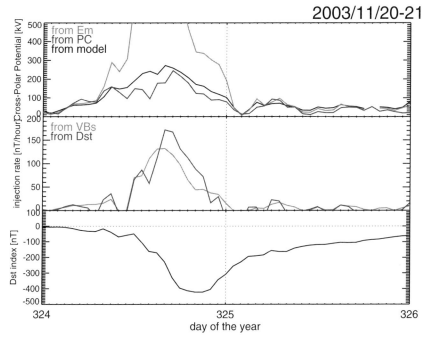

Fig. 1. (Top panel) Variations of cross-polar cap potential estimated from three independent methods during the super storm event on 20–21 November 2003. (Middle panel) Variations of injection rate estimated from two independent methods. (Bottom panel) Variations of Dst index.

In general, Q estimated from the solar wind parameters and the Dst index shows a good correspondence. However, a difference appears near the peak time of Q. The peak time of Q from the solar wind data is 2 h ahead of those from Dst.

One of the possibilities to cause the difference occurred during the first half of the main phase is the difference of solar wind parameter observed at ACE and at the Earth. OMNI-2 data is based on ACE observation during the event of our interest. However, almost the same magnetic field variations are observed at WIND, which is located at the tailward side of the Earth's magnetosphere at $(X, Y, Z) = (-211, -399, -11)$ in GSM coordinate system. This suggests that there is no difference of the solar wind parameters observed at ACE and at the Earth.

Next, we examine the properties of the magnetic field in the magnetosheath during this event by using the GOES satellites data when the satellite is in the magnetosheath. In this event, GOES 10 and 12 is in the

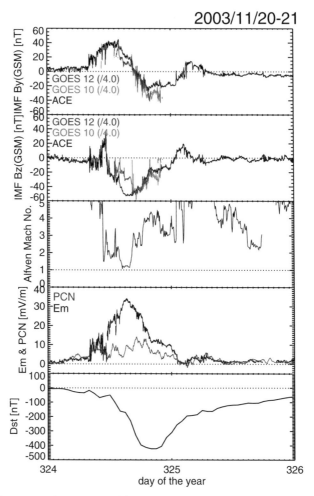

Fig. 2. Comparison between the ACE observations of the solar wind parameters and GOES observations of magnetosheath magnetic field during the magnetopause crossing period. (Top panel) GSM-By component of the magnetic field observed by ACE, and GOES 10 and 12; (second panel) GSM-Bz component of the magnetic field observed by ACE and GOES 10 and 12; (third panel) Alfven Mach number estimated from ACE observation; (forth panel) merging electric field and PCN index; (bottom panel) Dst index.

dayside geosynchronous orbit during the main phase of the storm, and both satellites cross the magnetopause. Figure 2 shows the comparison between the ACE observations of the solar wind parameters and GOES observations of magnetosheath magnetic field during the magnetopause crossing period.

Note that the intensity of the magnetic field observed by GOES is already divided by 4, which is the nominal compression ratio at the bow shock. After 16:30 UT, magnetic field variations observed by ACE and those observed by GOES show good correspondence. On the contrary, before 16:30 UT, magnetic field intensity observed by GOES is weaker than that observed by ACE, especially for Bz component. During this time period, Alfven Mach number goes down to 2, which is shown in the third panel. It is found that the period of the weakening of CPCP and Q estimated from the solar wind parameters corresponds to the time period of low Alfven Mach number.

3. Summary

We have studied how the S–M–I coupling drives magnetospheric convection and the geomagnetic storm during a super storm event on 20–21 November 2003. To examine the variations of the CPCP and the injection rate of the ring current, we estimate those two physical quantities from three and two different methods, respectively. In general, CPCP and Q during this super storm event are well reproduced from solar wind parameters and geomagnetic indices. CPCP estimated from the solar wind parameters based on Siscoe–Hill model is consistent with those estimated from the PCN index. It is clarified that the magnetospheric convection is highly saturated during this event. On the contrary, the variations of the geomagnetic storm represented by the Dst index develops linearly even during this extreme event. Q estimated from the solar wind parameters corresponds with those estimated from Dst. However, during the first half of the storm main phase, both CPCP and Q estimated from geomagnetic indices is less than those estimated from the solar wind parameters. The intensity of the magnetic field in the magnetosheath is small relative to those in the second half of the storm main phase because of the decreasing of Alfven Mach number below 2 within the intense magnetic cloud. Although the compression ratio at the bow shock goes down during the low Alfven Mach number, the intensity of the electric field conserves before and after the bow shock. This suggests that the efficiency of S–M–I coupling tend to be low during the low Alfven Mach number period. The S–M–I coupling controlled by the solar wind density during the low Alfven Mach number period was suggested by the previous studies.[10,11]

We conclude that the S–M–I coupling drives the magnetospheric convection and the geomagnetic storm variation in different manners even in the extreme conditions, and the efficiency of S–M–I coupling decreases when the Alfven Mach number is less than 2.

Acknowledgments

We thank the World Data Center for Geomagentism, Kyoto and the Danish Meteorological Institute for providing the Dst and the PCN index, respectively. We also thank the ACE instrument teams (SWEPAM and MAG) and the ACE Science Center for providing the ACE data. The OMNI data were obtained from the GSFC/SPDF OMNIWeb interface at http://omniweb.gsfc.nasa.gov.

References

1. T. P. O'Brien and R. McPherron, *J. Geophys. Res.* **105** (2000) 7707.
2. T. Nagatsuma, *Geophys. Res. Lett.* **29** (2002) 1422.
3. N. Gopalswamy, S. Yashiro, Y. Liu, G. Michalek, A. Vourlidas, M. L. Kaiser and R. A. Howard, *J. Geophys. Res.* **110** (2005) A09S15.
4. W. J. Burke, D. R. Weimer and N. C. Maynard, *J. Geophys. Res.* **104** (1999) 9989.
5. J. R. Kan and L. C. Lee, *Geophys. Res. Lett.* **6** (1979) 577.
6. G. L. Siscoe, G. M. Erickson, B. U. O. Sonnerup, N. C. Maynard, J. A. Schoendorf, K. D. Siebert, D. R. Weimer, W. W. White and G. R. Wilson, *J. Geophys. Res.* **107** (2002) 1075.
7. T. Nagatsuma, *J. Geophys. Res.* **111** (2006) A0902.
8. O. A. Troshichev, H. Hayakawa, A. Matsuoka, T. Mukai and K. Tsuruda, *J. Geophys. Res.* **101** (1996) 13,429.
9. J. H. King and N. E. Papitashvili, *J. Geophys. Res.* **110** (2005) A02209.
10. R. E. Lopez, M. Wiltberger, S. Hernandez and J. G. Lyon, *Geophys. Res. Lett.* **31** (2004) L08804.
11. R. Kataoka, D. H. Fairfield, D. G. Sibeck, L. Rastatter, M.-C. Fok, T. Nagatsuma and Y. Ebihara, *Geophys Res. Lett.* **32** (2005) L21108.

Advances in Geosciences
Vol. 14: Solar Terrestrial (2007)
Eds. Marc Duldig *et al.*
© World Scientific Publishing Company

LOW-LATITUDE E-REGION QUASI-PERIODIC ECHOES STUDIED USING LONG-TERM RADAR OBSERVATIONS OVER GADANKI

N. VENKATESWARA RAO and A. K. PATRA

National Atmospheric Research Laboratory
Tirupati, India

S. VIJAYA BHASKARA RAO

Department of Physics, Sri Venkateswara University
Tirupati, India

In this paper, we show various types of E-region quasi-periodic (QP) echoes based on a large database gathered using the Gadanki radar observations. The QP structures are found to have periods from tens of seconds to a few minutes. The echoes often found to be embedded in descending echoing region, show similar periods occurring in discrete echoing regions, and large height extent. Also, QP echoes are found to occur for duration as large as 10 h. Based on large database, we find that the QP echoes occurrence is much larger than its mid-latitude counterpart, an aspect that was not known earlier. Nighttime occurrence shows maximum in summer (45%), minimum in winter (14%), and moderate in equinoxes (26%–32%). The daytime occurrence shows again maximum in summer (58%) and nearly similar in other seasons (26.5% in March equinox, 32% in September equinox, 25.5% in winter). Both daytime and nighttime occurrences thus indicate that the QP echoes prefer to occur in summer. These aspects are discussed in the light of current understanding of QP echoes.

1. Introduction

One of the aspects of the ionospheric E-region irregularities revealed by radar observations that created an active area of scientific research is the quasi-periodic (QP) occurrence of E-region echoes. First observed in the mid-latitudes,[1] they are now known to occur at low latitudes as well.[2,3] These echoes usually display sloping striations with quasi-periodicity of 2–15 min in the range–time–intensity (RTI) or height–time–intensity (HTI) display as shown in Plate 2 of Yamamato *et al.*[1] Although low-latitude E-region phenomenology is different from that of mid-latitudes and also the magnetic dip angle for low-latitude is small in contrast to mid-latitude, surprisingly the basic characteristics of QP echoes (in terms of appearance of striations and period) at low latitudes closely resemble that of the mid-latitudes.

Although some of the basic features of the QP echoes from Gadanki radar observations have been reported,[4-6] they were based on limited observations. Much of the observations were episodic in nature and thus they did not reveal all possible features as well as their occurrence frequency so that a comparative study with the mid-latitude could be done. This is important considering the fact that all the theories developed to understand the QP echoes are based on mid-latitude observations, and the physical understanding have been developed keeping the mid-latitude E-region and magnetic field geometry in mind. Thus, the low latitude QP echoes have also been evaluated based on the mid-latitude processes. In this context, detailed characterization of low-latitude QP structures and their occurrence statistics would help understanding the physical processes responsible for their occurrence. Toward this goal, we have gathered a large amount of data on E-region field-aligned irregularities (FAI) using the Gadanki radar covering all local times and seasons spanning over a period of more than $2^{1}/_{2}$ years (June 2004–March 2007). This dataset allows us to examine a large variety of QP echoes occurring at different local times and heights. In this paper, we present various features of QP echoes and their occurrence statistics in a comprehensive way and discuss the observed features in the light of current understanding on the origin of these echoes.

2. A Brief Description About Experiment

Observations reported here were made using the Mesosphere–Stratosphere–Troposphere (MST) radar located at Gadanki (13.5°N, 79.2°E, magnetic latitude 6.4°N), which is a coherent pulse Doppler radar operating at 53 MHz with peak power-aperture product of 3×10^{10} W m^2 and half power beam width of 3°.[7] To study the fine temporal and spatial structures of the E-region FAI, experiments providing high temporal and spatial resolution observations were made for 5–7 days every alternate month during June 2004–March 2007. The experiments were conducted for 75 days and 88 nights. In order to detect the E-region FAI, the main lobe of the antenna pattern was positioned at an angle of 13° off-zenith due magnetic north that satisfies orthogonality condition to magnetic field at the E-region. Important radar parameters used for the experiments are

Table 1.   Radar specifications and parameters used
for the observations.

Location: Gadanki (13.5°N, 79.2°E, 6.4°N mag. lat)
Frequency: 53 MHz
Peak Power-Aperture product: 3×10^{10} W m^2
Antenna gain: 36 dB
Beam (3 dB full width): 3°
Beam direction: 13°N
Receiver bandwidth: 1.7 MHz
Inter-pulse Period (IPP): 2 ms
Pulse width: 16 μs with complementary code
Baud length: 1 μs
No. of coherent integrations: 4
No. of FFT points: 128

given in Table 1. The radar was operated with a coded pulse of 16 μs width and baud length of 1 μs, inter-pulse period of 2 ms, and echoes were sampled for 250 range bins covering the height range of 90–127.5 km. Echoes from four complementary pulses were added and 128 such samples for each range bin were used to compute power spectrum resulting in 250 Doppler spectra. Thus, to generate one set of power spectrum data it takes 1.024 s (2 ms $\times$ 4 $\times$ 128). Time taken for online computation, data transfer, and the radar controller to set up various subsystems for the next set of experiment is slightly less than 3 s. With the parameters given in Table 1, we obtained Doppler power spectrum with range resolution of 150 m, time resolution of 4 s (1.024 s plus $\sim$3 s), unambiguous velocity limit of ± 177 m s^{-1}, and velocity resolution of 2.76 m s^{-1}. The Doppler power spectra were stored for off-line processing to compute the moments. In off-line, first three lower order moments were computed to obtain total signal power, mean Doppler, and variance. Then, signal-to-noise ratio (SNR), mean phase velocity of the irregularities, and spectral width (2 $\times$ square root of variance) were computed. SNR computation involves noise power reckoned over the entire Doppler window of 125 Hz.

3. Observational Results

During the period of June 2004–March 2007, E-region FAI were found to display wide varieties of structures. Coming to the QP echoes, on which this paper focuses, they were found to vary greatly in terms of their appearance in HTI maps, their periods, and slopes of the striations. The periods, the height extents, the temporal extents of the striations, and the duration of

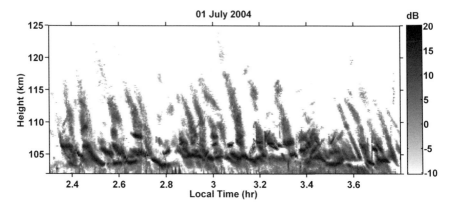

Fig. 1.   Height–time–intensity map of QP echoes observed on 1 July 2004.

event differ very much from event to event. In the following, we present various features of QP echoes and their statistics.

3.1. QP structures with different periods

Figure 1 shows a HTI map representing SNR of the radar echoes observed during nighttime on 1 July 2004. Range to height conversion has been made by multiplying the range with cos 13°, where 13° is the zenith angle of the radar beam. It may be mentioned that for the north bearing of the Gadanki radar observations, the difference between echo range and height is 2.5 km at 100 km altitude. QP echoing regions are found to extend to an altitude of 117 km in general and at times to altitude as high as 125 km. The largest striation is found to be 20 km. In this case, echo striations seem to be nearly continuous with height. Structures display mostly negative slope or no slope in this case. The strongest signal in this case is located in the undulated horizontal structure located at ∼105 km. All structures above 110 km are discrete in time irrespective of slope or strength. As evident, several periods can be discerned. While the dominant periods lie in the range of 4–7 min, period as short as 2.5 min also exists. While the periods shown in this example are very common in the Gadanki dataset, which could be categorized as conventional QP echoes, shorter than these periods are also observed quite frequently.

While the QP echoes with period more than the Brunt–Vaisala period (which is ∼5 min in the E-region) can be due to gravity waves, slightly shorter periods (say, more than 3 min) could be accounted for if the gravity waves are Doppler-shifted by background wind. However, for periods <3 min, there will be serious problems with the gravity-wave-related plasma structures since unrealistically large wind would be required. On the other hand, there are structures with periods of few tens of seconds[8] which when translated into spatial structures by multiplying with realistic plasma drift resemble closely the kilometer-scale gradient drift wave structures.[9] These are the zonally propagating primary plasma waves generated by the gradient drift instability with the free energy coming from a suitable combination of vertical electron density gradient and zonal drift.

Figure 2 shows an example of nighttime observations depicting short-period QP structures. These observations were made on 5 June 2005. Note that the periods are 1–2 min in this case. These structures resemble closely to those reported earlier by Patra and Rao.[8] They reported structures with periods of a few tens of seconds and attributed the structures to kilometer-scale gradient drift waves. These observations and those reported earlier thus indicate that structures with periods of tens of seconds to a few minutes could occur in the low-latitude E-region.

Short-period QP structures have been observed at lower altitudes as well. Figure 3 shows an example of short-period QP echoes observed in the evening of 18 February 2006. The dominant periods are 1–2 min, and the slopes of the structures are either negative or close to zero. Note that these structures are embedded in descending echoing region, and the descending echoing region comes down to an altitude of 89 km. Also, the lower echoing region while descending shows undulation with period

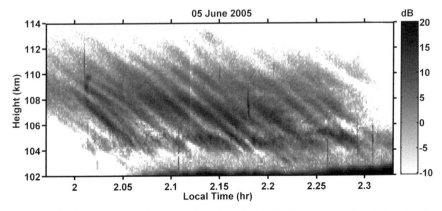

Fig. 2.   Height–time–intensity map of short-period QP echoes observed on 5 June 2005.

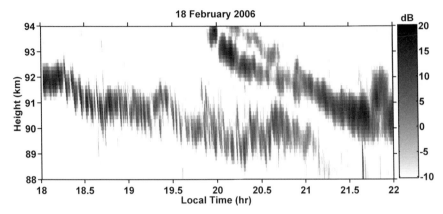

Fig. 3.   Height–time–intensity map of short-period QP echoes in descending echoing region observed in the evening of 18 February 2006.

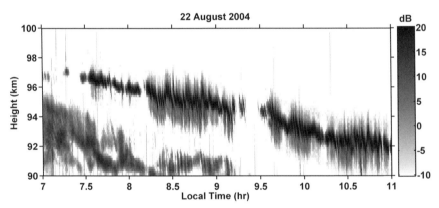

Fig. 4.   Height–time–intensity map of daytime QP echoes observed on 22 August 2004. Note the descending trend of the echoing region.

of ~30 min. The structures in such low altitudes call for an interpretation since the periods are similar to those of gradient drift waves but the region is highly collisional. On the other hand, the periods are too small to be considered in terms of gravity waves. These structures are not only observed during nighttime, but also during the day.

Figure 4 shows an example of daytime echoing event that shows structures in a descending echoing region occurring at lower E-region (<97 km). These observations correspond to 22 August 2004. Again in this

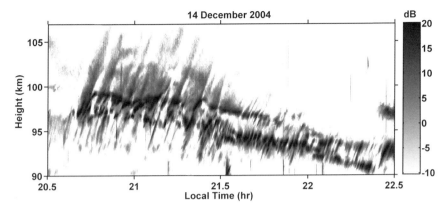

Fig. 5. Height–time–intensity map of positive-sloped QP echoes observed on 14 December 2004.

example, the dominant periods are 1–2 min with negative slopes, quite similar to those observed on 18 February 2006. In this case also, the descending echoing region in which these structures are embedded shows some undulation.

Although QP echoes mostly show negative slopes in the HTI maps, QP echoes having positive slopes have also been observed on some occasions. Their occurrence frequency, however, is negligibly small as compared to that of negative sloped structures. Figure 5 shows the QP echoes observed on 14 December 2004 with positive sloped striations during 20:35–22:25 LT. The periods are in the range of 2–5 min. Interestingly, the echoing region in which these structures are embedded is found to descend gradually with time, a feature that appears to be common irrespective of the slopes of the structures. Also note that initially the height extent of the striations is ∼10 km, which decreases to ∼3 km at a later part.

3.2. Multiple echoing region with similar periodic structures

An interesting feature of QP echoes over Gadanki that is quite common in our observations is the simultaneous occurrence of QP echoes in two or more distinctly different height regions with similar periodicity. Figure 6 shows an example of QP echoes observed in the evening hours of 24 February

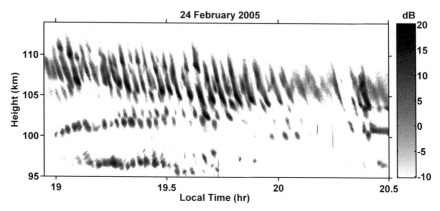

Fig. 6.   Height–time–intensity map of QP echoes observed on 24 February 2005 showing multiple echoing regions with similar periodicity.

2005. As many as three distinct echoing regions, all showing nearly similar periodicity, can be noted. The periods are 3–5 min, which fall in the commonly occurring periods of QP echoes at Gadanki. Again in this case also, the top echoing region is found to have a descending trend. QP echoes having similar periods occurring in discrete echoing regions raise important questions in regard to their generation mechanism, since they involve different altitudes, where the efficiency of a particular mechanism could vary remarkably. For example, in this case, the height region of interest is 96–115 km.

3.3. *Long duration of QP echoes in descending echoing region*

Figure 7 shows HTI map observed on 21–22 February 2005. QP echoes are clearly present right from the beginning (i.e. 19 LT), and such structures are found to keep on appearing throughout the night (~10 h) but with varying periods and at different heights. Note that the echoing region which occurred at ~112 km in the beginning descends to ~100 km. There are other echoing regions: one occurring after 22:30 LT at lower altitudes and is found to descend with time and another occurring after 24 LT at higher altitudes. The high altitude echoes, however, are weaker than those occurring below. The QP echoes in this example have periods 2–6 min. The important aspect to note from this example is that the descending echoing region shows features similar to those of tidal winds, and QP structures are present while it is descending.

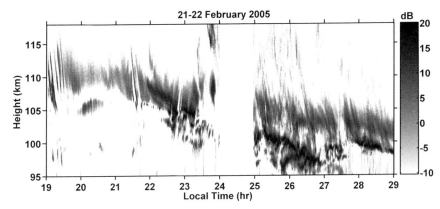

Fig. 7. Height–time–intensity map observed on 21–22 February 2005 showing the QP echoes to occur for long duration.

3.4. *Occurrence statistics*

High-resolution observations of the E-region FAI were made for 75 days and 88 nights during 2004–2007. Similar experiments were also made earlier but only during nighttime. Such observations were made for 59 nights. Out of 147 nights of radar observations, we have observed QP echoes for 83 nights (i.e. for 56% nights). During 147 nights, experiments were conducted for 1410 h and QP echoes were observed for a total period of 437 h (i.e. during 31% of the time in an overall sense). For daytime observations, out of 75 days, we have observed QP echoes for 46 days (i.e. 61% days). During the 75 days, experiments were conducted for 383 h and QP echoes were observed for a total period of 127 h.

Occurrence statistics of QP echoes for each season, providing total number of observational hours, total hours of QP echoes observed, and probability of occurrence (%), is given in Table 2. Number of nights of radar observations and number of nights on which QP echoes were observed in each season are also mentioned (in the bracket). Number of days and total hours of QP occurrence both clearly show that QP echoes have a strong seasonal preference during nighttime: maximum in summer (45%), minimum in winter (14%), and moderate in equinoxes (26%–32%). The daytime occurrence also shows a strong preference in summer (58%) like the nighttime QP echoes, but in other seasons their occurrences are nearly similar (26.5% in March equinox, 32% in September equinox, 25.5% in winter). Especially, in winter, the daytime QP echo occurrence is not as low as it is during nighttime.

Table 2. Statistics of QP echo occurrence.

Season	Nighttime			Daytime		
	Total observational hours (No. of nights)	Total hours of QP echoes observed (No. of nights QP echoes observed)	Probability of QP echo occurrence (%)	Total observational hours (No. of days)	Total hours of QP echoes observed (No. of days QP echoes observed)	Probability of QP echo occurrence (%)
March Equinox	285 (28)	93 (19)	32	119 (24)	31.5 (12)	26.5
Summer	425 (44)	192 (27)	45	69 (12)	40 (10)	58
September Equinox	440 (48)	116 (25)	26	110 (21)	35 (13)	32
Winter	260 (27)	37 (12)	14	85 (18)	20.5 (11)	25.5

4. Summary and Discussion

Based on the observations described above and those reported earlier, the important aspects of QP echoes that need to be addressed in terms of their generation mechanism are as follows:

(1) The QP structures occur at any altitude between 88 and 140 km and display a wide range of periods starting from a few tens of seconds to a few minutes.
(2) They occur both during the day and night, but daytime occurrence is confined to altitudes below 100 km and periods less than 5 min.
(3) They are often found embedded in descending echoing regions and can last for a couple of hours (2–10 h).
(4) The structures sometimes are found to occur at multiple discrete altitudes with similar periods and sometimes show continuous striations with the height extent of the striations as high as 20 km.
(5) Statistical results show that nighttime QP echoes occur for 14%–45% of the time and daytime QP echoes occur 25%–58% depending on the season. The highest occurrence rate is found in summer.
(6) Occurrence statistics of QP echoes over Gadanki are much higher than those of mid-latitudes.

4.1. *Morphology and occurrence statistics*

We have shown that low-latitude E-region QP echoes have structures with periods well below and above the Brunt Vaisala period with slopes mostly negative. Positive-sloped structures seldom occur over Gadanki. The height extent of the echoes is mostly below 120 km, occasionally extending to 135 km.[10] While the QP echoes extending to high altitude at mid-latitudes at times could be due to the ambiguity of the range to height conversion due to the mid-latitude magnetic field geometry, this ambiguity is only 2.5 km for the present observations due to low zenith angle of the radar beam.

On the other hand, we have shown simultaneous presence of QP echoes with similar period occurring at discrete altitudes. Importantly, the height extent in which they occur encompasses collision-dominated lower E-region also. Venkateswara Rao *et al.*[10] have reported simultaneous occurrence of undulation at lower altitude and discrete striations at higher altitudes. We will discuss this aspect later while dealing with the generation mechanism.

We have shown that the QP structures are often embedded in descending echoing region. QP echoes embedded in descending echoing region have also been reported from low-latitude locations, Piura[3] and Kototabang.[11] Such descending echoing regions have also been reported from mid-latitude locations of Shigaraki in Japan and Chung-Li in Taiwan.[12,13] These descending echoing regions are reminiscent of tidal ion layers. We may recall that the tidal ion layers are observed regularly using the Arecibo incoherent scatter radar and have been shown to be controlled by tidal wind fields.[14] Over and above these, gravity wave-like temporal modulation has been observed quite often in the Arecibo data.[14]

Coming to the occurrence of QP structures at Gadanki, for the nighttime QP echoes, we find that they occur 45% of the time in summer, 32% of the time in March equinox, 26% of the time in September equinox, and 14 % of the time in winter. For the daytime QP echoes we find that they are 58% of the time in summer, 26.5% of the time in March equinox, 32% of the time in September equinox, and 25.5% of the time in winter. Thus, the QP echoes show a strong seasonal variability with maximum occurrence being in summer. Interestingly, the daytime QP echo occurrence in winter is more than its nighttime counterpart.

For the nighttime QP echo occurrence at mid-latitudes, some statistics have already been reported and can be compared with those of Gadanki. Tsunoda et al.[15] based on observations from Tanegashima in Japan showed that echoes were observed during 48% of the time while QP echoes were observed only during 19% of the time. Hysell and Burcham[16] also provided statistics based on observations made from Clemson during the pre-midnight hours (20:00 LT, which is of 4 h duration) during 1–26 August 1998. They found that echoes were detected during 70% of the time and except for one echoing event, all echoes were QP in nature. Due to the operational hours used by them, the results need to be normalized for comparison. Normalizing these observations with those reported by Tsunoda et al.[15] we can roughly estimate the occurrence at Clemson to be ~23% and are comparable to that reported as 19% by Tsunoda et al.[15] Thus, the occurrence statistics of QP echoes over Gadanki are much higher than those of mid-latitudes, especially in summer. This finding clearly indicates that the governing factors manifesting QP echoes are often met at low latitudes, and possibly more frequent than at mid-latitudes, which is the most significant result not known earlier. On the other hand, from the dip equator, there exist no report on QP echoes of the type reported from low- and middle-latitudes except the one published by Woodman and Chau,[17] which was associated with a spectacular spread F event.

This clearly suggests that low-latitude E-region may be more conducive for manifesting QP echoes than their dip equator and mid-latitude counterparts.

As far as seasonal variation of the QP echoes over Gadanki are concerned, they are very much consistent with the seasonal variation of tidal wind activity in the Indian low-latitude region reported earlier based on meteor radar observations.[18] Gravity wave activities in the mesospheric temperature obtained using the Gadanki lidar observations also show quite similar variability.[19]

For the daytime QP structures, we have shown that the periods are <5 min and are generally lower than those of nighttime. The dominant periods are 1–3 min. Interestingly, their occurrence statistics is more than the nighttime. It may be noted that the daytime echoes are confined to altitudes below 100 km, and thus are related to collisional E-region. For the collisional region, such a high occurrence rate raise important question in terms of generation mechanism. We will discuss this aspect later while dealing with the physical processes leading to their generation.

4.2. *Physical processes responsible for QP echo formation*

Having discussed the morphological features, statistics, and seasonal dependence of the QP echoes at low latitudes, now we turn to the problem related to the physical processes responsible for their occurrence. So far the theoretical works aiming to explain various features of QP echoes have been concentrated addressing the mid-latitude observations. Few theories have been proposed to account for the mid-latitude QP echoes. They are: (1) Modulation of sporadic E (Es) layers by meridionally propagating gravity waves[20,21]; (2) Kelvin–Helmholtz instability (KHI)[22]; (3) Plasma blob and polarization processes[23]; and (4) Direction-dependent Es layer instability.[24] Although a clear picture is yet to emerge, the potential mechanisms and driving forces have been identified very well for mid-latitudes.

As far as the gravity wave mechanism[20,21] is concerned, it does not seem to be efficient for low latitudes, where magnetic field lines are nearly horizontal. Es structures generated by the meridionally propagating gravity waves will be prone to field-line shorting. Choudhary *et al.*[4] have also examined this mechanism from the background wind point of view to Doppler shift the intrinsic period of gravity waves to the observed ones, and found that unrealistically large wind is required for the applicability of this mechanism for Gadanki.

In regard to the direction-dependent Es layer instability,[21] Choudhary *et al.*[4] argued that it is not suitable for low latitudes since it requires the Es layers to be oriented near-vertical, which is unrealistic.

As far as the applicability of KHI[22] at low latitudes is concerned, it has been shown[4] that it is a potential mechanism, provided the necessary wind shear is present. The arguments in favor of this mechanism, however, were based on limited dataset. Based on the vast dataset, we find that many features cannot be explained based on KHI.[25] For example, the long duration of QP echoes cannot be explained based on KHI. Larsen[20] based on the simulation studies of KHI made by Fritts *et al.*[26] showed that the timescale for the buildup and restabilization of the shear flow is $\sim$30 min, which means that if the shear instability is the seed for the gradient drift instability, QP event should be observed for about 30 min at a time, and this could repeat as long as the shear flow repeats. Thus, the long duration of QP events observed at Gadanki are not consistent with the KHI.

In regard to the plasma blob and associated polarization processes[23] are concerned, this is quite plausible at low latitude. In a recent paper,[25] we have proposed a mechanism based on the zonal component of gravity wave winds. In that study, we have considered the zonal component of tidal winds to form Es layers and the zonal component of gravity wave winds to form plasma blob structures. The fact that the QP echoes are embedded in descending echoing region having features of tidal wind field seems to suggest the role of tidal neutral wind. We proposed that although tidal winds directly do not manifest QP echoes, they provide prerequisite conditions for the Es layers to form. The Es layers then turn QP plasma blobs presumably by the action of wind shear associated with gravity waves and get polarized to manifest QP echoes. On the other hand, the gravity wave winds at times providing unstable wind shear condition also could lead to KHI manifesting the QP echoes. Considering that the zonal component of gravity wave wind plays an important role in plasma structuring, which when unstable leads to QP echoes, the short periods observed frequently in our observations can be accounted for by the Doppler shifting of the intrinsic frequency by the background wind.

Similar arguments have been put forward by Mathews[14] to account for the high-altitude mid-latitude QP echoes observed by the MU radar.[1] His argument was based on the similarity observed between the Arecibo incoherent scatter radar maps of E-region electron density layers/structures and mid-latitude observations of E-region echoes revealed by the MU radar.

Recent coordinated observations made using Arecibo radar and coherent backscatter radar of the same ionospheric volume have shown coexistence of plasma blobs and QP echoes,[27] supporting this viewpoint.

4.3. *Unresolved issues and conclusion*

For the QP echoes occurring at higher altitudes while both gravity wave wind forming plasma blob and KHI mechanisms are equally plausible, at lower altitudes, the plasma blob related mechanism would not be efficient due to low efficiency of the winds forming plasma blobs there due to large collision frequency. For such echoes, KHI based on neutral wind shear driven instability providing seed plasma structures could be a potential mechanism. In such a case, even shear associated with the gravity wave winds could be a viable option. An important aspect of QP echoes that requires some concern for interpretation in terms of KHI is that some QP events last for a couple of hours. For the short duration QP echoes irrespective of height, the KHI will continue to be good possibility. However, for long-lasting QP event at lower altitudes, this seems to pose difficulty based on the existing knowledge of KHI.[22] This needs further investigations. Another point that needs some attention is the very low period (tens of seconds) observed in Gadanki observations.[28] The timescales are close to those of gradient drift plasma waves. While for the higher altitudes, this possibility remains intact, for collision-dominated lower E-region, the applicability of gradient drift instability at long wavelength of the order of a kilometer needs further investigation since their growth rate is known to be low there.

In conclusion, we find that the low-latitude QP echoes show a large variety of echo structures with the period varying from tens of seconds to a few minutes. All these periods cannot be accounted for by one mechanism. While the KHI and zonal component of gravity waves through plasma blob formation could account for the periods above $\sim$1 min, the short period (tens of seconds) does not seem to be accounted for by these mechanisms. Also we have shown that tidal winds play important role in providing conducive condition. It appears that gradient drift instability of some sort operating at long wavelength need to be considered. At this stage, however, it is not clear as to how these waves would grow at lower altitudes. Further investigations having concurrent measurements of electron density, electric field, and winds are necessary to understand the wide variety of low-latitude QP echoes.

Acknowledgments

The authors wholeheartedly appreciate the NARL technical staff for their dedicated efforts for making the observations reported here. One of the authors (NVR) is grateful to NARL for providing fellowship to him to carry out his research.

References

1. M. Yamamoto, S. Fukao, R. F. Woodman, T. Ogawa, T. Tsuda and S. Kato, *J. Geophys. Res.* **96** (1991) 15,943.
2. R. K. Choudhary and K. K. Mahajan, *J. Geophys. Res.* **104** (1999) 2613.
3. J. L. Chau and R. F. Woodman, *Geophys. Res. Lett.* **26** (1999) 2167.
4. R. K. Choudhary, J.-P. St.-Maurice, L. M. Kagan and K. K. Mahajan, *J. Geophys. Res.* **110** (2005) A08303, doi: 10.1029/2004JA010987.
5. C. J. Pan and P. B. Rao, *Geophys. Res. Lett.* **29** (2002) 1530, 10.1029/2001GL014331.
6. A. K. Patra, S. Sripathi, P. B. Rao and R. K. Choudhary, *Ann. Geophys.* **24** (2006) 1–9.
7. P. B. Rao, A. R. Jain, P. Kishore, P. Balamuralidhar, S. H. Damle and G. Viswanathan, *Radio Sci.* **30** (1995) 1125.
8. A. K Patra and P. B. Rao, *J. Geophys. Res.* **104** (1999) 24,667.
9. E. Kudeki, D. T. Farley and B. J. Fejer, *J. Geophys. Res.* **9** (1982) 684.
10. N. Venkateswara Rao, A. K. Patra, T. K. Pant and S. V. B. Rao, *J. Geophys. Res.* **113** (2008) A07312, doi: 10.1029/2007JA012830.
11. A. K. Patra, T. Yokoyama, M. Yamamoto, T. Nakamura, T. Tsuda and S. Fukao, *J. Geophys. Res.* **112** (2007) A01301, doi: 10.1029/2006JA011825.
12. M. Yamamoto, S. Fukao, T. Ogawa, T. Tsuda and S. Kato, *J. Atmos. Terr. Phys.* **54** (1992) 769.
13. C. J. Pan, C. H. Liu, J. Roettger, S. Y. Su and J. Y. Liu, *Geophys. Res. Lett.* **25** (1998) 1805, 10.1029/98GL00470.
14. J. D. Mathews, *J. Atmos. Solar Terr. Phys.* **60** (1998) 413.
15. R. T. Tsunoda, S. Fukao, M. Yamamoto and T. Hamasaki, *Geophys. Res. Lett.* **25** (1998) 1765.
16. D. L. Hysell and J. D. Burcham, *J. Geophys. Res.* **104** (1999) 4361.
17. R. F. Woodman and J. L. Chau, *Geophys. Res. Lett.* **28** (2001) 207.
18. C. R. Reddi and G. Ramkumar, *J. Atmos. Solar Terr. Phys.* **59** (1997) 1757.
19. G. Ramkumar, T. M. Antonita, Y. Bhavani Kumar, H. Venkata Kumar and D. Narayana Rao, *Ann. Geophys.* **24** (2006) 2471.
20. R. F. Woodman, M. Yamamoto and S. Fukao, *Geophys. Res. Lett.* **18** (1991) 1197.
21. R. T. Tsunoda, S. Fukao and M. Yamamoto, *Radio Sci.* **29** (1994) 349.
22. M. F. Larsen, *J. Geophys. Res.* **105** (2000) 24,931.
23. T. Maruyama, S. Fukao and M. Yamamoto, *Radio Sci.* **35** (2000) 1155.
24. R. B. Cosgrove and R. T. Tsunoda, *Geophys. Res. Lett.* **29**(18) (2002) 1864, doi: 10.1029/2002GL014669.

25. N. Venkateswara Rao, A. K. Patra and S. V. B. Rao, *J. Geophysics. Res.* **113** (2008) A03309, doi: 10.1029/2007JA012574.

26. D. C. Fritts, T. L. Palmer, Ø. Andreassen and I. Lie, *J. Atmos. Sci.* **53** (1996) 3173.

27. D. L. Hysell, J. Drexler, E. B. Shume, J. L. Chau, D. E. Scipion, M. Vlasov, R. Cuevas and C. Heinselman, *Ann. Geophys.* **25** (2007) 457.

28. A. K. Patra, S. Sripathi, V. Siva Kumar and P. B. Rao, *Geophys. Res. Lett.* **29** (2002) 1499, 10.1029/2001GL013340.

Advances in Geosciences
Vol. 14: Solar Terrestrial (2007)
Eds. Marc Duldig *et al.*
© World Scientific Publishing Company

MIRROR MODES OBSERVED WITH CLUSTER IN THE EARTH'S MAGNETOSHEATH: STATISTICAL STUDY AND IMF/SOLAR WIND DEPENDENCE

V. GÉNOT*, E. BUDNIK, C. JACQUEY and I. DANDOURAS

CESR-CNRS-UPS, Observatoire Midi-Pyrénées
Toulouse, France
**vincent.genot@cesr.fr*

E. LUCEK

The Blackett Laboratory, Imperial College
London, UK

We present a statistical analysis of five years of Cluster mission data in the magnetosheath. Our primary focus is to exhibit the spatial distribution of mirror mode events. The automatized detection is based on Minimum Variance Analysis and the amplitude of events. The results are displayed in the GIPM reference frame to enable comparison with a previous similar study using ISEE-1 data. These results are compared favorably with each other and with the studies focusing on Jupiter's and Saturn's magnetosheaths. We further analyze the dependence of the mirror events with solar wind parameters and Interplanetary Magnetic Field (IMF) orientation. We notably reveal that the occurrence of mirror modes is relatively more probable during the periods of time when the IMF is not following the common Parker spiral orientation.

1. Introduction

Mirror mode structures in space plasma continue to fuel many new works in spite of an already vast literature on the subject. Hereafter, some of the main characteristics of the mirror mode are reviewed to explain this abundant and active research activity. Its ubiquitousness: in environments of the Earth,[1,2] Jupiter,[4-6] Saturn,[7] Io wake,[8] the comet Halley,[9] solar wind,[10] ICME,[11] in the heliosheath,[13,14] and even in turbulent galaxy clusters[15]; its peculiarity: a nonpropagating mode, it is known as a fluid mode, obtained as one of the modes of anisotropic MHD, although it requires a kinetic treatment (including Landau damping, see also Ref. 16) to properly determine the growth rate, because its transverse length scale extends to the order of the proton Larmor radius (see the discussion in Ref. 17); its potential existence at large to small length scales[18]; its role in plasma transport in particular near the magnetopause[19]; all of these features make the mirror mode and mirror instability fundamental objects of study in plasma physics.

The threshold required for mirror instability to develop in a proton–electron plasma is given by[20,21]

$$\beta_{p\perp}\left(\frac{T_{p\perp}}{T_{p\parallel}}-1\right)+\beta_{e\perp}\left(\frac{T_{e\perp}}{T_{e\parallel}}-1\right)>1+\frac{\left(\frac{T_{p\perp}}{T_{p\parallel}}-\frac{T_{e\perp}}{T_{e\parallel}}\right)^2}{2\left(\frac{1}{\beta_{p\parallel}}+\frac{1}{\beta_{e\parallel}}\right)},\qquad(1)$$

where T is the temperature, $\beta=2\mu_0 nkT/B^2$, subscripts $\perp$ and $\parallel$ stand for the directions with respect to the ambient magnetic field B, subscripts e and p are for electrons and protons, respectively. For isotropic cold electrons, this condition reduces to

$$\beta_{p\perp}\left(\frac{T_{p\perp}}{T_{p\parallel}}-1\right)>1.\qquad(2)$$

Close to the Earth, the best location for this condition to be met is in the magnetosheath (the temperature anisotropy and β values are indeed relatively large). Any observational study related to the mirror mode thus required a proper definition of the boundaries of the magnetosheath, the bow shock and the magnetopause, as well as a proper way of localization in this region. Long-term operating missions such as ISEE and now Cluster enable the possibility of comprehensive statistical studies to reveal key dependence in the occurrence of wave modes with local and remotely controlling parameters. This is the main task of the present paper which addresses the occurrence and dependence issues of the mirror mode in the present Cluster context (see also Refs. 1 and 22). The chapter is organized as follows. Section 2 presents the data and the way magnetosheath is identified. In Sec. 3, the characterizing methods we employ for mirror modes are discussed. Section 4 exposes results of the statistical analysis with Cluster compared with the results obtained from 10 years of ISEE-1 data,[2,3] whereas Sec. 5 is concerned with solar wind and IMF dependencies. A summary of the main findings of the study concludes the chapter in Sec. 6.

2. Data

Five years of the Cluster mission are considered (01/02/2001 to 31/12/2005). Cluster 1 magnetic field (FGM[23]) and on-board calculated ion moments (from the HIA experiment on the CIS instrument[24]) data are used at 4 s resolution. We also employ ACE plasma and IMF data to determine the magnetopause and bow shock positions using models (as described below).

A web-based version of the statistical analysis tool developed at CDPP (the French Plasma Physics Data Center) and used in this study is available at the URL: cdpp-amda.cesr.fr. Access is granted upon request (mail to amda@cesr.fr).

2.1. *Magnetosheath identification*

The first step of our analysis is to determine whether Cluster is located in the magnetosheath. Data are analyzed by 5 min window: an iterative delay procedure is applied to obtain associated solar wind and IMF parameters from ACE. Bow shock[3] and magnetopause[25] models are computed from these parameters, and Cluster is then identified as "in" or "out" of the magnetosheath. Data will be displayed in the GIPM (geocentric interplanetary medium) reference frame first introduced by Ref. 46 and detailed in Ref. 3. This will enable consistent comparison with ISEE-1 results which are displayed in this frame in Ref. 3. For the sake of clarity we reproduce here the definition of this reference frame.

The x-axis $\mathbf{e}_x$ is antiparallel to the upstream solar wind velocity vector $\mathbf{V}$ in the reference frame moving with the planet (with components V_x, V_y, V_z in the GSE reference frame); the direction of the GIPM y-axis is determined by the IMF vector $\mathbf{B}$ (with components B_x, B_y, B_z in the GSE reference frame):

$$\mathbf{e}_y = \mathrm{sgn}(\mathbf{B} \cdot \mathbf{e}_x) \cdot (-\mathbf{B} + (\mathbf{B} \cdot \mathbf{e}_x)\mathbf{e}_x)/|\mathbf{B} - (\mathbf{B} \cdot \mathbf{e}_x)\mathbf{e}_x|. \tag{3}$$

With such a definition, an IMF field line lies in the second and fourth quadrants of the GIPM xy plane. Zenith angle (θ) and clock angle (ϕ) are defined by

$$\theta = \arccos\left(\frac{\mathbf{r} \cdot \mathbf{e}_x}{r}\right), \tag{4}$$

$$\phi = \arctan\left(\frac{\mathbf{r} \cdot \mathbf{e}_z}{\mathbf{r} \cdot \mathbf{e}_y}\right), \tag{5}$$

where **r** is a vector in the GSE frame. The clock angle is measured perpendicular to the solar wind direction. For an average IMF direction (following the Parker spiral) $-90° < \phi < 90°$ corresponds to the dusk magnetosheath side and $90° < \phi < 270°$ is on the dawn side.

In the following, we define our GIPM reference frame by using a 20 min averaged (and shifted) IMF vector centered on the selected magnetosheath event.

2.2. *Fractional distance in the magnetosheath*

We will also make use of the fractional distance F introduced by Ref. 3 to normalize event positions in the magnetosheath:

$$F = \frac{r - r_{MP}}{r_{BS} - r_{MP}}, \tag{6}$$

where r is the Cluster geocentric distance, r_{MP} is the geocentric distance to the magnetopause (which is a function of the zenith angle, the solar wind ram pressure, and IMF B_z), and r_{BS} is the geocentric distance to the bow shock (which is a function of the zenith and clock angles, the upstream Alfvén and Mach numbers, and the angle between the solar wind velocity and IMF vectors). Therefore, $F = 0$ at the magnetopause and $F = 1$ at the bow shock.

3. Mirror Mode Characterization

Identification of mirror mode events has been a long-standing problem, because: (1) slow mode and mirror mode have both anti-correlated magnetic and density fluctuations; (2) mirror mode and ion cyclotron mode both grow on temperature anisotropy ($T_\perp > T_\parallel$). However, in the magnetosheath, it has been shown[26] and observed[27] that due to the presence of heavier ions (mainly helium), mirror instability dominates for $\beta > 1$ which is the most common situation, a condition confirmed by recent simulations[28] ($\beta_\parallel \geq 0.35$).

Along time, and with improving tools at hand, different methods have been developed to discriminate low-frequency modes and among them the mirror mode. These include (1) transport ratio,[29,30] (2) minimum variance analysis,[2] (3) 2- and 4-spacecraft methods,[16,31–33] (4) 90° B/V_z phase difference.[34] A complete review is out of the scope of this chapter, but the interested reader may refer to Ref. 35 for more details.

Multi-spacecraft studies have shown that mirror mode structures are elongated along a direction making a small angle with the ambient magnetic field.[1,33,36]

3.1. *Identification of mirror-like structures*

The magnetic field variations associated with mirror modes are almost linearly polarized parallel to the main field direction. They may be of large amplitude (a few 10%) of sinusoidal but also spiky up and down shapes as we shall discuss in Sec. 4.4. From these characteristics, a criterion has been established which follows closely those used by Refs. 2 and 37:

- Linear polarization with field variation oriented close to the ambient magnetic field: the angle between the maximum variance direction and mean magnetic field vector is smaller than 20°.
- Relatively large amplitude: the variance of the field is larger than 10%.

In order to perform statistical survey over 5 years of data we had to employ relatively low resolution data (4 s), which limits the lower sampled mirror event size to 8 s. From a 2 month survey with high-resolution Cluster data, Ref. 37 found that mirror events were distributed as a bell-shaped distribution with 98% of events falling into the 4 s–24 s interval and with a mean of 12 s. This shows that our data set is undersampled as it misses events in the 4–8 s length which corresponds to the events with the smaller spatial scale (of the order of 10 local Larmor radii). However, short duration mirror event does not automatically translate into short length scale event as the spacecraft velocity and geometry effects have to be taken into account. Therefore, the undersampling of our data set is not one-to-one equivalent to a bias toward large length scale mirror events. Nevertheless, the way event scales are affecting statistics still remains to be studied with high-resolution data.

The above criterion is applied to all 5 min magnetosheath intervals obtained in the first step. The mean magnetic field is calculated on 10 min window and the Minimum Variance Analysis[38] (MVA), and variance are performed on 5 min windows. The sensitivity of the results has been tested against the variation of these time windows, and it revealed no major difference. At this stage, we do not make any restriction on the plasma parameter values as we are interested in mirror-like structures appearing above as well as below the linear mirror instability threshold given by Eq. (2). No constraint has been imposed on the eigenvalues λ resulting from the MVA. Indeed, as noted by Ref. 16 (from a small set of mirror events though), mirror modes are more commonly observed as elliptically than linearly polarized as predicted by linear theory. It has been checked that restraining our data set to linear events (for instance, with the condition: $\lambda_{\mathrm{int}}/\lambda_{\mathrm{max}} \leq 0.2$ and $\lambda_{\mathrm{min}}/\lambda_{\mathrm{int}} \geq 0.3$) does not significantly alter the conclusions drawn in the rest of the chapter.

Let us note that Ref. 2 supplement their criterion with a condition on the symmetry of the structures which essentially selects magnetic depressions. Therefore, "dip" or "hole" mirror modes only are discussed in Ref. 2. This makes a significant difference between ISEE-1 and Cluster data sets which is discussed at the end of Sec. 4.

Automatic detection of data patterns is hard to be perfect, and it is therefore possible that compressional structures (like quasi-perpendicular shocks) other than mirror modes may be selected by our algorithm. It is however difficult to evaluate the proportion of misinterpreted events. In order to limit this occurrence, we shall use error bars or plot data with only sufficient statistical significance.

3.2. *Mirror condition*

In order to qualify the plasma state with respect to the mirror instability, we define the distance to threshold by

$$C_M = \beta_{p\perp} \left(\frac{T_{p\perp}}{T_{p\parallel}} - 1 \right). \tag{7}$$

From Eq. (2), $C_M < 1$ ($C_M > 1$) corresponds to mirror (un)stable plasma while $C_M = 1$ denotes marginal stability for which the mirror growth rate is zero.

4. Cluster Statistics and Comparison with ISEE-1 Data

The comparison exposed in this section is based on the results obtained by the algorithm presented in Sec. 3. Table 1 lists the range of orbital parameters associated with ISEE-1 and Cluster missions. Time resolutions of both these missions are identical.

In Fig. 1, the total number of 5 min magnetosheath crossings is displayed in the zenith angle — fractional distance plane in bins $\Delta\theta \times \Delta F = 5° \times 0.05$. Events are integrated over all ϕ angles. The maximum number of crossings per bin is close to 10,000. Cluster orbital configuration leads

Table 1. Ranges of parameters relative to the ISEE-1 mission[2,3] and Cluster mission (this study).

Mission	ISEE-1	Cluster
Time range	10 y	5 y
Time resolution	4 s	4 s
Fractional distance (F) range	0–1	0–1
Zenith angle (θ) range	20°–100°	0°–150°

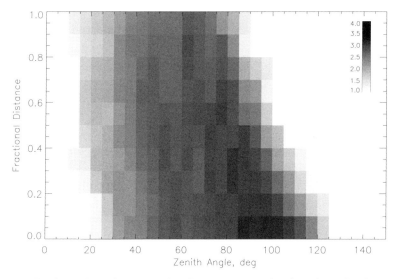

Fig. 1. Total number of magnetosheath crossings in the (zenith angle, fractional distance) plane. The gray scale is logarithmic.

to a larger coverage close to the magnetopause and for high latitudes. In comparison, ISEE-1 orbits covered a slightly reduced zenith angle range (see Table 1). Plotting data in a different plane, namely, the clock angle–fractional distance plane (in bins $\Delta\phi \times \Delta F = 5° \times 0.1$ integrated over $0° \leq \theta \leq 150°$; figure not shown) illustrates that all regions of the magnetosheath are correctly sampled with a majority of events close to the magnetopause.

4.1. Occurrence frequency

The distribution of mirror mode events (i.e. 5 min intervals which satisfy criteria of Sec. 3.1) is displayed in Fig. 2 (zenith angle–fractional distance

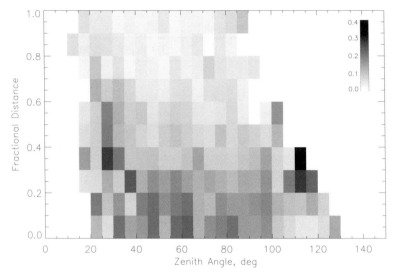

Fig. 2. Relative number of mirror mode events in the (zenith angle, fractional distance) plane. The gray scale codes the occurrence frequency and bins with less than five magnetosheath crossings have been rejected.

plane) and Fig. 3 (clock angle–fractional distance plane); bins with less than five magnetosheath crossings have been rejected in order to perform a statistically meaningful normalization. For this latter representation, let us note that data from dawn (dusk) side of the magnetosheath are displayed on the left (right) part of the plot. Equivalently, this corresponds to the quasi-parallel shock (quasi-perpendicular) region (see Sec. 2.1 and Ref. 3 for more details on the GIPM reference frame). The number of events is divided by the total number of magnetosheath crossings to reveal the relative number of mirror events (or occurrence frequency). This shows that mirror events are more likely to occur close to the magnetopause for all zenith angles and for smaller and smaller angles for increasing distance from the magnetopause. The dark bin at $\theta = 112.5°$ and $F = 0.35$ is a statistical artifact. There is a dawn/dusk asymmetry with more events occurring in the dusk sector, which is also the region connected to the quasi-perpendicular shock after which mirror modes are mostly expected. Indeed larger temperature anisotropies are generally encountered behind perpendicular shocks rather than parallel ones due to a sharper transition from solar wind to magnetosheath plasmas (see also the theoretical work in Ref. 12). These results agree with those of Refs. 2 and 3. Although our data representation is more pixelized than that in Ref. 3, occurrence

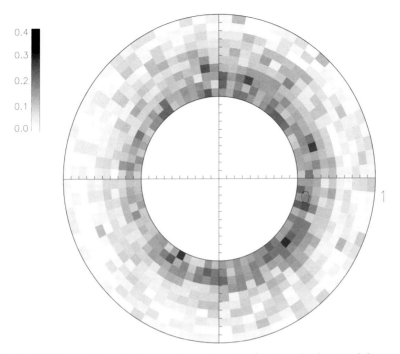

Fig. 3.   Relative number of mirror mode events in the (clock angle, fractional distance) plane. The gray scale codes the occurrence frequency, and bins with less than five magnetosheath crossings have been rejected.

frequencies cannot be one-to-one compared as normalization has been applied differently. The present findings are also consistent with the observations from Equator-S data which show that mirror modes are mostly encountered in the inner magnetosheath region.[39]

4.2. *Amplitude distribution*

The amplitude distribution of mirror mode events $\delta B/B$ is displayed in Fig. 4 in a representation equivalent to Fig. 3; bins with less than five mirror events have been rejected in order for the statistics to be meaningful. There is a tendency to observe larger events in the middle magnetosheath although the distribution is scarce close to the shock. Also, the average intensity of mirror structures is generally larger in the morning magnetosheath ($90° < \phi < 270°$) compared to the evening magnetosheath region ($-90° < \phi < 90°$). It is more precise in the pre-dawn quadrant that mirror amplitudes are larger ($90° \leq \phi \leq 135°$) in close agreement with the ISEE-1 results.

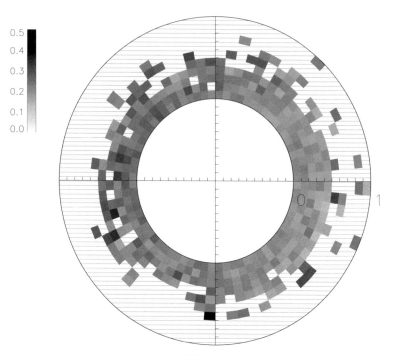

Fig. 4. Relative mirror mode amplitude ($\delta B/B$) in the (clock angle, fractional distance) plane. The gray scale codes the magnitude of the magnetic field perturbation and bins with less than five mirror events have been rejected.

Reference 3 argues that there may be a bias effect due to the high level of turbulence behind the parallel bow shock which hides low amplitude mirror fluctuations. A similar effect may affect our data, or if this is a real physical feature, the process behind it remains not fully understood.

The results from Cluster observations are however consistent with those obtained from ISEE-1 reported in Refs. 2 and 3: the maximum value for $\delta B/B$ is comparable ($\sim$0.5 in both studies). They also compare favorably with the results from the studies on giant planets. At Saturn (Ref. 7), it was found that the amplitude and wavelength of fluctuations tend to increase with increasing distance from the quasi-perpendicular bow shock, except close to the magnetopause in the plasma depletion layer. At Jupiter, Erdös and Balogh[4] studied the statistical properties of mirror mode depressions observed by Ulysses and found that the amplitude of the fluctuations was decreasing when approaching the bow shock. Naturally, mirror waves need time to grow from their supposed origin at the shock, and as they are convected toward the planet simultaneously, larger fluctuations are seen away from the shock. Close to the magnetopause (1) the free energy contained in the anisotropy may have been consumed by instabilities, (2) the wave growth may have saturated, or (3) the plasma flow has been deviated (near the plasma depletion layer), which explained why amplitude does not peak close to the magnetopause.

4.3. Growth rate

Following Hasegawa,[40] Liu *et al.*[11] derived an expression for the maximum growth rate of the mirror instability (normalized to the proton cyclotron frequency Ω_p):

$$\frac{\gamma_m}{\Omega_p} = \frac{1}{\sqrt{12\pi\beta_\perp}} \left(\frac{T_{p\perp}}{T_{p\parallel}}\right)^{-3/2} \left(\beta_\perp \left(\frac{T_{p\perp}}{T_{p\parallel}} - 1\right) - 1\right)^2, \qquad (8)$$

for the conditions $k_\parallel \ll k_\perp$ in the long wavelength limit and $C_M > 1$ (see Eq. (7)).

Based on a numerical evaluation of the full kinetic dispersion relation, this maximum growth rate is $\gamma_m/\Omega_p \simeq 0.02$ for a plasma containing 5%–10% alpha particles, $T_\perp/T_\parallel = 1.5$, $T_{\alpha\parallel} = 4T_{p\parallel}$, and $\beta_\perp = 4$ (see Refs. 26 and 47). With these plasma parameters, the above equation also yields $\gamma_m/\Omega_p \simeq 0.02$. Let us note that since this value is significantly larger than the maximum growth rate of the proton cyclotron instability, mirror waves can grow faster.

By using Eq. (8), therefore selecting mirror events with $C_M > 1$ only in our analysis, maximum growth rate is evaluated to be below the $\gamma_m/\Omega_p = 0.1$ level, whereas 77% of events are below the $\gamma_m/\Omega_p = 0.01$ level; the mean value is 0.008. These values are consistent with the observations and simulations showing that the magnetosheath plasma is mostly in a marginal state with respect to the mirror instability.[28]

Using a model of plasma flowlines and data from the ISEE-1 spacecraft, Tatrallyay and Erdös[41] showed that the growth rate values are in the range $0.002\,\mathrm{s}^{-1} < \gamma < 0.0035\,\mathrm{s}^{-1}$, which is almost an order of magnitude smaller than the value calculated by Ref. 26. These growth rates are not maximum values, but are computed from the evolution of $\delta B/B$ along flow lines. Reference 41 proposes several reasons for this discrepancy, one of which being that the source of the mirror fluctuations may not be at the bow shock, but at various locations more deep inside the magnetosheath.

Also, the linear analysis performed by Gary *et al.*[26] might not be applicable to large-amplitude nonlinear fluctuations. However, we note that the authors restrained their data set to magnetic dips (or holes) only. Recent works[17,43] have shown that such magnetic configurations generally (1) correspond to a late evolutionary stage of mirror modes (nonlinear regime) and (2) are observed in mirror stable plasma ($C_M < 1$), both conditions in which application of Eq. (8) or linear theory is not appropriate.

4.4. Discussion: Differences between ISEE-1 and Cluster data sets

As discussed above, the algorithm in Ref. 2 retained only magnetic depressions, whereas our present results concern all magnetic shapes. This is not an innocuous remark. Indeed, the shape of mirror modes had recently attracted attention both from the theoretical[17,42] and observational[6,37,43] points of view. These works gave new insight in the physics governing the evolution of the mirror instability. Previous studies had identified that mirror structures come in different shapes,[16,39] but it is only recently that it was revealed that the shape was controlled by the distance to the threshold C_M. It was concluded that deep holes are mainly due to a bistability process (enabling the existence of mirror structures in mirror-stable plasma conditions, $C_M < 1$); moderate holes and peaks may be observed near threshold; and large peaks are nonlinearly saturated mirror mode structures far from threshold. The localization of these different structures in the magnetosheath showed that holes are preferentially observed close to the magnetopause, whereas peaks are more frequent in the middle magnetosheath. This spatial distribution is illustrated in Fig. 5 which shows the value of C_M in the (zenith angle, fractional distance) plane: maximum value, corresponding to peaks, are obtained in the middle magnetosheath whereas close to the magnetopause C_M is close or less than the one denoting the presence of holes. The proportion of events with $C_M > 1$ ($C_M < 1$) is 43% (57%). A more detailed description is out of the scope of the present chapter (the relation shape/C_M is exposed extensively in Refs. 37 and 43), but these remarks enable to pinpoint differences between our present analysis and the one based on ISEE-1. Reference 2 retained magnetic depressions only, mostly observed in mirror stable plasmas close to the magnetopause, whereas our analysis is more general in keeping all mirror-like structures without constraining plasma parameters. However, observations show that peaks are a minority (14% of events in Ref. 6,

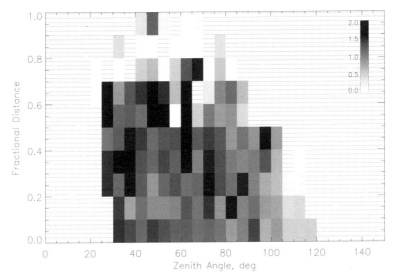

Fig. 5.   Distribution of the average distance to threshold (C_M) of the events in the (zenith angle–fractional distance) plane.

18.7% in Ref. 37 for 19% and 39.7% of hole structures, respectively) in mirror data sets; simulations also show that large mirror peaks only survive in seldom-encountered large β plasmas.[43] This may explain why results compare favorably between both studies despite differences in the initial data sets.

5. Relation with Solar Wind Parameters and IMF Orientation

We use ACE data and a solar wind–magnetosheath iterative delay algorithm to associate each magnetosheath event to the corresponding solar wind and IMF parameters. From the values given in Table 2, it is interesting to note that mean values for both kinds of events (mirror and nonmirror) are not significantly different, except for M_A. In this particular case, the striking difference led us to investigate this dependence into more details.

5.1. M_A dependence

In Fig. 6 the distribution of events with M_A bins (of width 1) is plotted: the dash line is for all magnetosheath of 5 min intervals, whereas the solid line

Table 2.   Mean values of solar wind and IMF parameters for
mirror and nonmirror types of events.

Type of events	Mirror	Nonmirror
Number of events	6363	57405
Alpha/proton density ratio	0.0453	0.0448
Ram pressure (nPa)	2.33	2.12
M_A ($M_A^2 = 4\pi\rho V^2/B^2$)	*10.91*	*8.17*
M_s ($M_s^2 = \rho V^2/\gamma p$)	8.68	8.49
γ_m/Ω_p	0.008	—

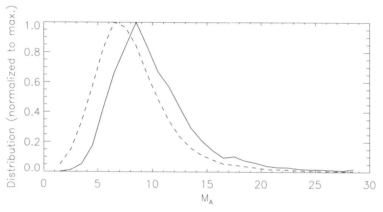

Fig. 6.   Distributions of mirror-like structures (solid line) and of all magnetosheath
crossings (dash line) as a function of M_A.

refers to identified mirror modes only, both normalized to their maximum
value. Distributions exhibit clear peaks with a larger most probable value
for mirror events ($\max(M_A) = 8.5$) than for nonmirror-associated traversals
($\max(M_A) = 6.5$). The average value of the distribution is also higher for
mirror events (10.91 to be compared to 8.17, see Table 2). The ratio between
these two distributions is plotted in Fig. 7 to reveal that the occurrence
frequency of mirror modes increases as a function of M_A until $M_A = 12$.
This increasing trend is the prominent feature of the figure as more than
80% of mirror events occur for $M_A \leq 12$.

As the distribution of mirror events with the ram pressure does not
show significant shift compared with the distribution of magnetosheath
crossings (and equivalently for the dependence with M_s; not shown,
see mean values in Table 2), the dependence on M_A translates into a
dependence on the IMF magnitude. Whereas the upstream (IMF) and

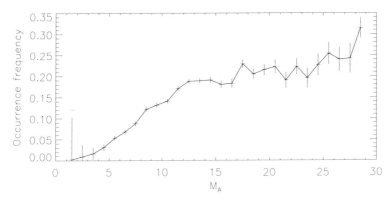

Fig. 7.   Occurrence frequency of mirror-like structures as a function of M_A. The error bars are proportional to $1/\sqrt{N}$ where N is the number of mirror events in each $\Delta M_A = 1$ bin ($N_{\min} = 1$ and $N_{\max} = 909$).

downstream (magnetosheath) magnetic fields are (positively) correlated, it is less clear for the temperature. Therefore, as mirror modes are favored by large $\beta (\propto 1/B^2)$ conditions, this magnetic field correlation can tentatively explain the variation with $M_A (\propto 1/B)$.

As solar wind perturbations acting on the global response of the magnetosphere are studied to exhibit geo-effectiveness, the solar wind parameters acting decisively on the growth of mirror modes have to be studied in more details, and this beyond the "simple" filtering effect of the bow shock which has to be taken into account evidently, as we discuss below.

5.2. *IMF orientation*

The occurrence of mirror mode signatures in relation with the IMF orientation has been analyzed. Mirror modes are usually more frequently observed behind quasi-perpendicular shocks (see Sec. 4.1), which for common orientation of the IMF (along the Parker spiral) corresponds to the dusk side. To construct Fig. 8, for each 5 min magnetosheath crossing, we recorded the associated IMF orientation and plotted the corresponding point in the (B_x, B_y) plane (in GSE coordinates). In this plane, most of the points are observed in the second and fourth quadrants along the $B_y = -B_x$ line, which corresponds to the 45° Parker spiral (proportions of events in quadrants 2–4 and 1–3 are 77% and 23%, respectively). When we retain mirror mode events only the picture is changed as the other two quadrants exhibit significantly more points, making the distribution appear like a

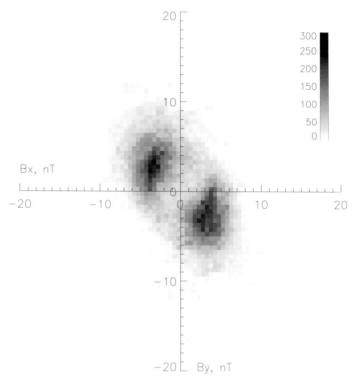

Fig. 8.   IMF in the XY_{GSE} plane for all magnetosheath crossings. The gray scale codes the number of events in each $0.5\,\mathrm{nT} \times 0.5\,\mathrm{nT}$ bin.

ring (proportions are then 70% and 30%). When one plots the relative number of events (Fig. 9) the result is even more striking as it appears clearly that this relative number is higher for IMF direction perpendicular to the average Parker spiral. This original result has also been observed with ISEE data (M. Tátrallyay, private communication). Indeed, ISEE data revealed that there were relatively more events at the time of nontypical IMF directions compared to the events when the IMF was closer to the Parker spiral direction: in about 75% of all observations, the IMF was in the Parker quadrants, whereas for about 25% of all observations the IMF was in the nontypical quadrants; but only about 70% of the selected mirror events were in the Parker quadrants, and about 30% of these events were in the nontypical quadrants. These values are remarkably consistent with those obtained with Cluster data. Therefore, according to these numbers the selected mirror events occurred more than 30% more frequently when

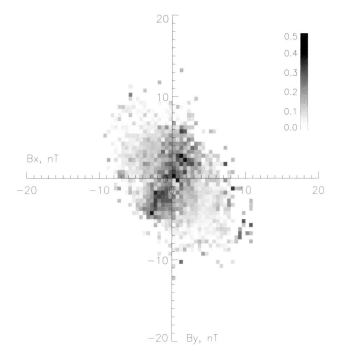

Fig. 9. IMF in the XY_{GSE} plane for mirror only events normalized to the total number of crossings. The gray scale codes the occurrence frequency, and bins with less than five crossings have been rejected.

the IMF is not in the Parker spiral direction (i.e. an increase from 23% to 30% in the Cluster case).

Similar plots with Cluster data in the (B_y, B_z) plane (not shown) exhibit a smaller increase of this occurrence frequency for non-Parker situations ($\sim$7% to be compared to the 30% above).

To understand fully the process behind this counter-intuitive observation, one needs to compute correctly the nature of the shock associated with each mirror event. This implies using a streamline model (derived from Ref. 44 for instance) or simulation results (like it is done in Ref. 6 using a model by Ref. 45) which is left for future work.

6. Summary and Conclusion

We have used 5 years of Cluster magnetosheath crossings to investigate the occurrence of fluctuations associated with the mirror instability.

No constrain was imposed on the plasma parameters, as it is recognized that mirror structures also exist below the mirror instability linear threshold (mainly in the form of magnetic holes[43]). Let us summarize the main findings of the chapter.

- There is a larger occurrence in the inner region of the magnetosheath, close to the magnetopause at larger zenith angle and closer to the middle of the sheath in the subsolar region.
- There is a dawn/dusk asymmetry with more events occurring in the dusk sector, i.e. behind quasi-perpendicular shocks.
- Mirror fluctuation amplitudes are larger in the middle magnetosheath (despite a small statistical coverage close to the shock).
- There is a dawn/dusk asymmetry with more large amplitude events occurring in the dawn sector, i.e. behind quasi-parallel shocks.
- There is no significant dependence with M_s or with the ram pressure of the solar wind.
- There is a clear dependence with M_A which, given the above remarks, translates into a dependence with the IMF amplitude B.
- There is a clear dependence with the IMF $(B_x, B_y)_{GSE}$ angle with relatively more mirror events (+30%), occurring at times when the IMF is not oriented along the Parker spiral.

Such conclusions could be interesting for integration in further modeling efforts. Indeed, clarifying the space of potentially influencing parameters is crucial for studying the evolution of the mirror instability. For instance, it is such an observational study on mirror structure shapes which paved the way to identify the bi-stability phenomenon which is behind the shape/localization mirror structure filtering in the magnetosheath. Improvements in the present analysis could imply the use of a plasma flow line model to relate properly the events to their originating bow shock localization. Regarding the influence of the solar wind, this analysis clearly opened new questions (role of IMF magnitude and orientation), which demand a parametric study to be properly addressed using numerical simulations.

Acknowledgments

This work was performed in the frame of the International Space Science Institute meetings of Team 80 "The effect of ULF turbulence and flow

chaotization on plasma energy and mass transfer at the magnetopause."
The authors thank the CDPP team members for their efforts in developing
the analysis tools used in this study. VG also thanks M. Tátrallyay,
P. Hellinger, T. Passot, and G. Belmont for fruitful discussions.

References

1. E. A. Lucek, M. W. Dunlop, T. S. Horbury, A. Balogh, P. Brown, P. Cargill, C. Carr, K.-H. Fornaçon, E. Georgescu and T. Oddy, *Ann. Geophys.* **19**(10) (2001) 1421.
2. M. Tátrallyay and G. Erdös, *Planet. Space Sci.* **53** (2005) 33.
3. M. I. Verigin, M. Tátrallyay, G. Erdös and G. A. Kotova, *Adv. Space Res.* **37** (2006) 515.
4. G. Erdös and A. Balogh, *J. Geophys. Res.* **101** (1996) 112.
5. N. André, G. Erdös and M. Dougherty, *Geophys. Res. Lett.* **29**(20) (2002) 1980, doi: 10.1029/2002GL015187.
6. S. P. Joy, M. G. Kivelson, R. J. Walker, K. K. Khurana, C. T. Russell and W. R. Paterson, *J. Geophys. Res.* **111**(A12) (2006) A12212.
7. M. B. Bavassano Cattaneo, C. Basile, G. Moreno and J. D. Richardson, *J. Geophys. Res.* **103** (1998) 11,961–11,972.
8. D. E. Huddleston, R. J. Strangeway, X. Blanco-Cano, C. T. Russell, M. G. Kivelson and K. K. Khurana, *J. Geophys. Res.* **104**(A8) (1999) 17,479–17,490, 10.1029/1999JA900195.
9. C. T. Russell, W. Riedler, K. Schwingenschuh and Y. Yeroshenko, *Geophys. Res. Lett.* **14**(6) (1987) 644–647.
10. D. Winterhalter, M. Neugebauer, B. E. Goldstein, E. J. Smith, S. J. Bame and A. Balogh, *J. Geophys. Res.* **99** (1994) 23,371–23,381.
11. Y. Liu, J. D. Richardson, J. W. Belcher, J. C. Kasper and R. M. Skoug, *J. Geophys. Res.* **111**(A9) (2006) A09108.
12. Y. Liu, J. D. Richardson, J. W. Belcher and J. C. Kasper, *Astrophys. J.* **659**(1) (2007) L65–L68.
13. L. F. Burlaga, N. F. Ness and M. H. Acũna, *Geophys. Res. Lett.* **33** (2006) L21106, doi: 10.1029/2006GL027276.
14. L. F. Burlaga, N. F. Ness and M. H. Acũna, *J. Geophys. Res.* **112** (2007) A07106, doi: 10.1029/2007JA012292.
15. A. A. Schekochihin, S. C. Cowley, R. M. Kulsrud, M. S. Rosin and T. Heinemann, Nonlinear growth of firehose and mirror fluctuations in turbulent galaxy-cluster plasmas, ArXiv e-prints (2007) 0709.3828.
16. V. Génot, S. J. Schwartz, C. Mazelle, M. Balikhin, M. Dunlop and T. M. Bauer, *J. Geophys. Res.* **106**(A10) (2001) 21,611–21,622, 10.1029/2000JA000457.
17. T. Passot, V. Ruban and P. L. Sulem, *Phys. Plasmas* **13** (2006) 102310.
18. F. Sahraoui, G. Belmont, L. Rezeau, N. Cornilleau-Wehrlin, J. L. Pincon and A. Balogh, *Phys. Rev. Lett.* **96**(7) (2006) 075002.

19. J. R. Johnson and C. Z. Cheng, *J. Geophys. Res.* **102**(A4) (1997) 7179–7190, 10.1029/96JA03949.

20. A. N. Hall, *J. Plasma Phys.* **21** (1979) 431.

21. P. Hellinger, *Phys. Plasmas* **14** (2007) 8.

22. O. D. Constantinescu, K.-H. Glassmeier, R. Treumann and K.-H. Fornacon, *Geophys. Res. Lett.* **30**(15) (2003) 1802, doi: 10.1029/2003GL017313.

23. A. Balogh *et al.*, *Ann. Geophys.* **19** (2001) 1207.

24. H. Rème *et al.*, *Ann. Geophys.* **19**(10) (2001) 1303–1354.

25. J.-H. Shue, J. K. Chao, H. C. Fu, C. T. Russell, P. Song, K. K. Khurana and H. J. Singer, *J. Geophys. Res.* **102**(5) (1997) 9497.

26. S. P. Gary, S. A. Fuselier and B. J. Anderson, *J. Geophys. Res.* **98**(A2) (1993) 1481–1488, 10.1029/92JA01844.

27. B. J. Anderson, S. A. Fuselier, S. P. Gary and R. E. Denton, *J. Geophys. Res.* **99**(A4) (1994) 5877–5892, 10.1029/93JA02827.

28. P. Trávníček, P. Hellinger, M. G. G. T. Taylor, C. P. Escoubet, I. Dandouras and E. Lucek, *Geophys. Res. Lett.* **34**(15) (2007) L15104, doi: 10.1029/2007GL029728.

29. P. Song, C. T. Russell and S. P. Gary, *J. Geophys. Res.* **99**(A4) (1994) 6011–6026, 10.1029/93JA03300.

30. R. E. Denton, S. P. Gary, B. J. Anderson, J. W. LaBelle and M. Lessard, *J. Geophys. Res.* **100**(A4) (1995) 5665–5680, 10.1029/94JA03024.

31. G. Chisham, S. J. Schwartz, M. A. Balikhin and M. W. Dunlop, *J. Geophys. Res.* **104**(A1) (1999) 437–448, 10.1029/1998JA900044.

32. M. A. Balikhin, O. A. Pokhotelov, S. N. Walker, E. Amata, M. Andre, M. Dunlop and H. S. C. K. Alleyne, *Geophys. Res. Lett.* **30**(10) (2003) 1508, doi: 10.1029/2003GL016918.

33. T. S. Horbury, E. A. Lucek, A. Balogh, I. Dandouras and H. Reme, *J. Geophys. Res.* **109** (2004) 9209, doi: 10.1029/2003JA010237.

34. C.-H. Lin, J. K. Chao, L. C. Lee, D. J. Wu, Y. Li, B. H. Wu and P. Song, *J. Geophys. Res.* **103**(A4) (1998) 6621–6632, 10.1029/97JA03474.

35. S. J. Schwartz, D. Burgess and J. J. Moses, *Ann. Geophysicae* **14** (1996) 1134–1150.

36. A. N. Fazakerley and D. J. Southwood, *Adv. Space Res.* **14**(7) (1994) 65.

37. J. Soucek, E. Lucek and I. Dandouras, *J. Geophys. Res.* **113** (2008) A04203, doi: 10.1029/2007JA012649.

38. B. U. O. Sonnerup and L. J. Cahill, Jr., *J. Geophys. Res.* **72** (1967) 171.

39. E. A. Lucek, M. W. Dunlop, A. Balogh, P. Cargill, W. Baumjohann, E. Georgescu, G. Haerendel and K.-H. Fornacon, *Ann. Geophys.* **17** (1999) 12.

40. A. Hasegawa, *Phys. Fluids* **12** (1969) 2642.

41. M. Tatrallyay and G. Erdös, *Planet. Space Sci.* **50** (2002) 593–599.

42. E. A. Kuznetsov, T. Passot and P. L. Sulem, *Phys. Rev. Lett.* **98** (2007) 235,003.

43. V. Génot, E. Budnik, P. Hellinger, T. Passot, G. Belmont, P. Trávníček, E. Lucek and I. Dandouras, *Annales Geophysicae* **27**(2) (2009) 601–615.

44. J. R. Spreiter and S. S. Stahara, *J. Geophys. Res.* **85** (1980) 6769.

45. T. Ogino, R. J. Walker and M. G. Kivelson, *J. Geophys. Res.* **103** (1998) 225.

46. J. W. Bieber and E. C. Stone, *Proceedings of Magnetospheric Boundary Layers Conference*, Alpbach, 11–15 June 1979, ESA SP-148, 1979, pp. 131–135.

47. D. Hubert, C. Lacombe, C. C. Harvey and M. Moncuquet, *J. Geophys. Res.* **103** (1998) 26,783.

Advances in Geosciences
Vol. 14: Solar Terrestrial (2007)
Eds. Marc Duldig et al.
© World Scientific Publishing Company

[10]Be CONCENTRATIONS IN SNOW AT LAW DOME, ANTARCTICA FOLLOWING THE 29 OCTOBER 2003 AND 20 JANUARY 2005 SOLAR COSMIC RAY EVENTS

J. B. PEDRO[*,†], A. M. SMITH[‡], M. L. DULDIG[§], A. R. KLEKOCIUK[§],
K. J. SIMON[‡], M. A. J. CURRAN[†,§], T. D. VAN OMMEN[†,§], D. A. FINK[‡],
V. I. MORGAN[†,§] and B. K. GALTON-FENZI[*,†]

*Institute of Antarctic and Southern Ocean Studies
University of Tasmania, Hobart, Tasmania 7005, Australia

†Antarctic Climate and Ecosystems CRC
Hobart, Tasmania 7005, Australia

‡Australian Nuclear Science and Technology Organisation
Menai, New South Wales 2234, Australia

§Australian Antarctic Division
Kingston, Tasmania 7050, Australia

Recent model calculations have attempted to quantify the contribution of major energetic solar cosmic ray (SCR) events to [10]Be production.[1,2] In this study we compare modeled [10]Be production by SCR events to measured [10]Be concentrations in a Law Dome snow pit record. The snow pit record spans 2.7 years, providing a quasi-monthly [10]Be sampling resolution which overlaps with the SCR events of 29 Oct 2003 and 20 Jan 2005. These events were calculated to increase monthly [10]Be production in the polar atmosphere ($>65°$ S geomagnetic latitude) by $\sim$60% and $\sim$120% above the GCR background, respectively.[2] A strong peak in [10]Be concentrations ($>4\sigma$ above the 2.7 y mean value) was observed $\sim$1 month after the 20 Jan 2005 event. By contrast, no signal in [10]Be concentrations was observed following the weaker 29 Oct 2003 series of events. The concentration of [10]Be in ice core records involves interplay between production, transport, and deposition processes. We used a particle dispersion model to assess vertical and meridional transport of aerosols from the lower stratosphere where SCR production of [10]Be is expected to occur, to the troposphere from where deposition to the ice sheet occurs. Model results suggested that a coherent SCR production signal could be transported to the troposphere within weeks to months following both SCR events. We argue that only the 20 Jan 2005 SCR event was observed in measured concentrations due to favorable atmospheric transport, relatively high production yield compared to the 29 Oct 2003 event, and a relatively high level of precipitation in the Law Dome region in the month following the event. This result encourages further examination of SCR signals in [10]Be ice core data.

1. Introduction

^{10}Be ($t_{1/2} = 1.5 \times 10^6$ y) is a cosmogenic radionuclide produced by the interaction of cosmic rays with the Earth's atmosphere. The majority of ^{10}Be is produced by spalation of O, N, and Ar atoms by galactic cosmic rays (GCRs) within the energy range 0.1–10 GeV.[3] Following production, ^{10}Be is rapidly scavenged by aerosols and transported with air masses. Some ^{10}Be is deposited and archived within the annual snow layers of the polar ice caps. Ice core records of ^{10}Be concentration sample this archive and may be used to reconstruct the history of the factors which control the flux of cosmic rays to Earth, principally variations in solar activity[4,5] and variations in the geomagnetic field strength.[6,7]

Reconstructions of solar activity from ice core ^{10}Be concentrations can provide information for assessing solar forcing of climate in the past and hence the role of solar forcing in the present climate change.[8,9] However, reconstruction of reliable solar activity records from ^{10}Be has been limited by poor understanding of the influence of climate processes on ^{10}Be transport and deposition to the polar ice core sites.[5,10,11] The challenge in interpreting ^{10}Be records is to separate variations in production rate in the atmosphere caused by changes in solar activity from variations caused by atmospheric mixing, transport pathways, depositional processes, and changes in the precipitation rate.

A useful test of ^{10}Be records as proxies of solar activity is comparison of the ice core record to the instrumental record of solar activity and cosmic ray flux (CRF). Satellite records of solar activity were initiated in the 1970s. Instrumental records of CRF began with the installation of ionization chambers in the 1930s and then longer term, more stable ground-based neutron monitors in the 1950s. However, few comparisons between measured ^{10}Be and instrumental data exist.[12–14]

High-resolution ^{10}Be records spanning the modern era are also required to answer questions regarding possible solar cosmic ray (SCR) contribution to ^{10}Be production. SCRs are energetic particles emitted by the Sun generally in the energy range 0.1–50 GeV. During a major SCR event, the flux of particles arriving at the Earth's atmosphere can increase by orders of magnitude.[1] Over decadal and longer timescales, the production of ^{10}Be by SCRs is probably small (1%–2%) compared to production by GCRs.[1] However, over annual and shorter timescales SCR events may lead to significant production of ^{10}Be.

Recent theoretical studies have been presented for SCR contributions to ^{10}Be production during the ~70 y record of instrumental SCR data.[1,2]

Usoskin *et al.*[1] argue that major events can cause up to a doubling in annual polar production of ^{10}Be. Webber *et al.*[2] suggest a smaller magnitude of SCR contribution, finding at most ~50% enhancement in polar production (for the large February 1956 event). The calculations of Webber *et al.*[2] may be more reliable as they use updated atmospheric yields based on latest cross sections and a broader spectral range of incident protons. Measurements of ^{10}Be concentration in polar ice cores for periods spanning major SCR events are required to test and compare with modeled production rates. Such a comparison requires consideration of the latitudinal gradient in ^{10}Be production rate and the poorly constrained influences of atmospheric transport and deposition processes on ^{10}Be fallout.

Since SCRs are generally of lower rigidity (momentum per unit charge) than GCRs, they largely have access restricted to the polar regions. Polar production rates are therefore much more sensitive to SCRs than global average production rates. For example, the February 1956 SCR event was calculated to cause a ~50% increase in polar production but only ~12% increase in global average production.[2] It follows that the sensitivity of polar ice core records to SCR signals is highly dependent on the extent of meridional mixing of ^{10}Be produced over the polar regions. There is evidence that meridional mixing of ^{10}Be produced over the poles is limited, for example, from general circulation model calculations[15] and from comparison of ice core ^{10}Be concentrations with the amplitude of the geomagnetic and solar variations in the past.[12,16]

The atmospheric lifetime of ^{10}Be produced in the stratosphere is several years, whereas in the troposphere ^{10}Be is washed out by precipitation processes within several weeks.[17] Stratospheric concentrations of ^{10}Be are therefore enhanced with respect to the troposphere, and the temporally and spatially variable processes of stratosphere–troposphere exchange are of high relevance to ^{10}Be fallout. In addition to mixing processes, local effects including precipitation rate and frequency also influence the ^{10}Be fallout.[5,11,15]

In November 2005, a 4 m deep snow pit was extracted at Law Dome Summit (the "DSS0506 snow pit"). The snow pit has time resolution and dating accuracy at the quasi-monthly level and spans 2.7 y (March 2003 to November 2005). The record overlaps with two major SCR events: 29 Oct 2003 and 20 Jan 2005. Here, we examine the DSS0506 snow pit for SCR signals from these events. We argue that high temporal resolution ice core records of ^{10}Be concentration for periods spanning major SCR events have potential to inform on both the magnitude of the SCR contribution to ^{10}Be production and the extent of latitudinal mixing of ^{10}Be. Through consideration of SCR production signals as "pulse" inputs of ^{10}Be to the atmosphere, information may also be obtained on atmospheric residence time and temporal variations in ^{10}Be transport.

2. Methods

2.1. Sample site and characteristics

Law Dome, East Antarctica (66° 46′ S, 112°48′ E) (Fig. 1) is well suited to the extraction of precisely dated and continuous chemical records in fine detail.[18,19] It experiences high snow accumulation (0.68 m y^{-1} ice equivalent) which provides for high temporal resolution records (quasi-monthly), low annual average wind speed (8.3 ms^{-1}), which minimizes surface disturbance and perennially low temperature (summer mean, $-12.6°C$) which precludes alteration of records by summer melt.[20]

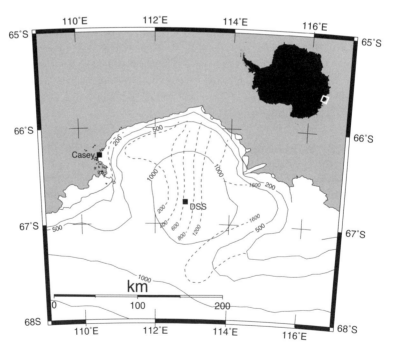

Fig. 1. Location of Dome Summit South (DSS) sample site, Law Dome, Antarctica with accumulation isopleths (mm ice equivalent, dashed lines) labeled.

2.2. Sample collection

The DSS0506 snow pit was excavated to a depth of 4.0 m in November 2005 near Law Dome Summit (1379 m asl, 66°46′ 11″ S, 112°48′ 41″ E). Forty 10 cm thick contiguous samples of approximately 1.5 kg mass were taken down the face of the snow pit by insertion of a 10 cm high × 20 cm×20 cm stainless steel scoop (following Pedro *et al.*[11]). Samples were immediately sealed in zip lock bags. To minimize contamination, all equipments were pre-rinsed with >18 MΩ cm^{-1} Milli-Q water, and personnel were equipped with dust masks and gloves.

2.3. Sample preparation and measurement

Sample preparation and accelerator mass spectrometer (AMS) measurements were carried out at the Australian National Tandem Accelerator for Applied Research (ANTARES) AMS using the techniques described in Child *et al.*[21] and Fink *et al.*[22] ^{10}Be concentrations were normalized to the National Institute of Standards (NIST) SRM 4325 ^{10}Be standard with an adjusted ratio of 3.02×10^{-11}.[23] Measurements of the NIST standard were reproducible within ±2%. Chemistry procedural blanks had very low values of ^{10}Be/^{9}Be ($<10 \times 10^{-15}$), indistinguishable from the carrier, indicating neither machine background nor chemistry processes introduced ^{10}Be at any significant level. After normalization to the NIST standard, error-weighted mean ^{10}Be/^{9}Be ratios ranged from $(210-966) \times 10^{-15}$, with overall errors of <4%.

The low ^{10}Be/^{9}Be ratios achieved in chemistry blanks from this study resulted from our use of a beryllium carrier derived from the mineral beryl which is very low in ^{10}Be, along with a step in our chemistry processing technique introduced specifically to remove boron (^{10}B) contamination from samples (^{10}B is a problematic isobaric interferent in AMS measurement of ^{10}Be). This technique involved fuming BeO from each sample with hydrofluoric acid in sulfuric acid to volatilize boron as BF$_3$ (K. Simon, in prep).

2.4. Sample dating

Fractionation of water isotopes during evaporation, transport, and precipitation processes leads to a well-established relationship between the oxygen isotope ratio (δ^{18}O) of precipitation and site temperature at high latitude sites.[24] This is the basis for climatic temperature reconstructions from ice cores and is clearly evident in strong seasonal cycles in $\delta^{18}O$ observed at Law Dome.[25] The clear summer maxima in $\delta^{18}O$ provide dating horizons for Law Dome ice core records.

$\delta^{18}O$ samples were taken from the bulk melt water of each 10 cm resolution ^{10}Be snow pit sample. The snow pit $\delta^{18}O$ record is therefore registered directly with the ^{10}Be record. The timescale was determined by picking summer peaks in $\delta^{18}O$, assisted by interlocking with $\delta^{18}O$ records from other recent ice cores from Law Dome Summit and by reference to absolute dating tie points. Figure 2 shows the DSS0506 snow pit $\delta^{18}O$ record with $\delta^{18}O$ records from two recent ice cores, the DSS0506 ice core (drilled 300 m away from the DSS0506 snow pit) and the DSS0405 ice core (drilled at the same site two years earlier). The three records are generally in good agreement, which allows confidence in dating the DSS0506 snow pit. Some differences occur in the records due to local influences on snow removal, ablation, and accumulation. The separation of the DSS0506 ice core record and the DSS0506 snow pit record over the first 30 cm (ice equivalent) is due to the compression of the topmost portion of the snow pack by the drilling process and does not reduce confidence in dating.

Experimental work shows that on average $\delta^{18}O$ peaks on 10 January at Law Dome,[25] although year-to-year variations in timing of precipitation and annual temperature maximum are acknowledged. Year boundaries are labeled in Fig. 2 by attribution of 10 January to the annual $\delta^{18}O$ maxima. Where there is disagreement between the three $\delta^{18}O$ records, a mean is taken. Firm tie points on the dating scale are the sample dates for the DSS0506 snow pit (8 Nov 2005) and the drilling date of the DSS0405 ice core (31 Oct 2004).

The dates of samples between summer peaks are interpolated assuming even snow accumulation over the course of a year. Evidence suggests that this assumption of uniformly distributed precipitation is valid when averaged over many years,[25] however in any given year, precipitation biases are seen.[19] The bottom date for the record is determined by interlocking the snow pit $\delta^{18}O$ record with the DSS0506 ice core $\delta^{18}O$ record. Our best estimate of error associated with dating of the snow pit record is ± 1 month, with dating confidence firmer in the vicinity of the 8 Nov 2005 and 31 Oct 2004 tie points.

Reproducibility of snow pit records extracted from Law Dome is expected to be good; prior studies found that trace chemical records extracted from multiple snow pits on Law Dome provide records which are in agreement over at least a 12 km transect.[26]

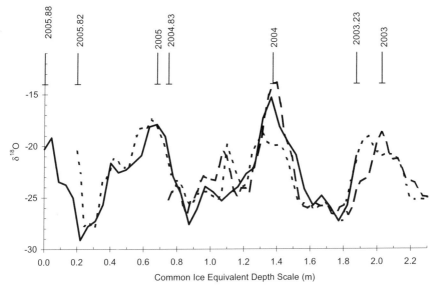

Fig. 2. Dating of the DSS0506 snow pit record was carried out using $\delta^{18}O$ with cross reference to two recent ice core records from Law Dome and registration against firm dating tie points. DSS0506 snow pit $\delta^{18}O$ record (solid line), DSS0506 ice core $\delta^{18}O$ record (dashed line), and DSS0405 ice core $\delta^{18}O$ record (dotted line). The interpolated year boundaries for 2005, 2004, and 2003 are labeled. Firm dating tie points are sampling dates of the DSS0506 snow pit (2005.88), DSS0506 ice core (2005.82), and DSS0405 ice core (2004.83). The bottom date of the DSS0506 snow pit is interpolated (2003.23). The date scale is expected to be accurate to ±1 month with more confidence close to dating tie points.

3. Results

3.1. ^{10}Be concentration and variability in the DSS0506 snow pit

The mean (and SD) of the concentration of ^{10}Be in the 40 snow pit samples was $(6.0 \pm 2.0) \times 10^3$ atoms g^{-1}. This value is similar to earlier measurements of ^{10}Be concentration in modern Law Dome ice, e.g. 5.6×10^3 atoms g^{-1} reported by Pedro et al.[11] and 7.6×10^3 atoms g^{-1} reported by Smith et al.[27] Figure 3(a) shows variation in ^{10}Be concentration with time for the snow pit record. The record spans March 2003 to December 2005 with dating accuracy estimated at ±1 month. The range of concentrations varies by up to fourfold throughout the record from 2.7 to 12.6×10^3 atoms g^{-1} (with 4% analytical errors). Concentrations are generally higher during the late

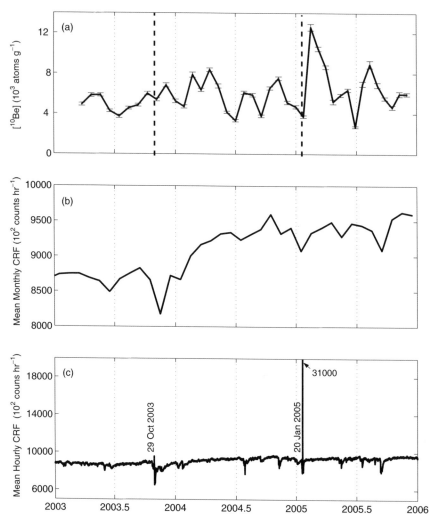

Fig. 3. (a) ^{10}Be concentration in the DSS0506 snow pit showing timing of 29 Oct 2003 and 20 Jan 2005 SCR events (dashed lines); (b) McMurdo monthly mean CRF[31]; (c) McMurdo hourly mean CRF[31] with the 29 Oct 2003 and 20 Jan 2005 Ground Level Enhancements labeled.

austral summer (summer) and lower during the austral winter (winter). This is in agreement with a prior snow pit study at Law Dome which found a late summer maximum in ^{10}Be concentration.[11] Recurrence of the late summer maximum in ^{10}Be concentrations suggests that atmospheric transport is conducive to increased ^{10}Be fallout at this time of year; prior studies have suggested that an enhancement in stratosphere to troposphere exchange may be implicated.[11,28,29]

Over time periods of less than 1 year, meteorological processes are expected to dominate variability in ^{10}Be concentrations.[10] However, some secondary influence by production rate changes in the atmosphere may also be imprinted on the record. The most prominent feature in the snow pit record is the strong increase in ^{10}Be concentration during February of 2005. A possible SCR origin for this peak is discussed in Sec. 3.3.

3.2. ^{10}Be concentration and monthly mean cosmic ray flux

Solar modulation of the GCR flux arriving at the Earth's atmosphere over the course of a 11-yr Schwabe cycle is calculated to cause variation in global ^{10}Be production between 15% and 50%.[30,31] At polar latitudes ^{10}Be production is more sensitive to variation in the GCR flux; it has been calculated that production changes over recent Schwabe cycles at polar latitudes are on average close to 80%.[2] Here we examine the possible influence of shorter term (monthly) variations in the CRF as recorded by ground-based neutron monitors on ^{10}Be concentrations in ice.

Figure 3(b) shows monthly mean CRF data[32] recorded at the McMurdo neutron monitor (the closest monitor to Law Dome with available data). This may be compared to ^{10}Be concentration in the snow pit in Fig. 3(a). No correlation is apparent between the monthly CRF data and quasi-monthly ^{10}Be concentrations, either direct or delayed. Over the period of the snow pit record, mean monthly CRF varies by at most ~5%, and such relatively small variations in CRF cannot explain the large variations in ^{10}Be concentration observed in the snow pit. Our data support the contention that meteorological processes are the principal cause of variation in ice core concentrations of ^{10}Be over time periods of $<1\,y$ (e.g. Beer[10]). To our knowledge this is the first comparison of monthly resolution CRF data to ^{10}Be concentrations at similar resolution from a polar ice core site.

We note that the positive trend in CRF over the period spanned by the record is matched by an increasing trend in ^{10}Be concentration. This is consistent with the Schwabe cycle influence on CRF and ^{10}Be production in the atmosphere, but the level of variability in the ^{10}Be concentration precludes affirmation of a casual link.

3.3. ^{10}Be concentration and solar cosmic ray events

^{10}Be production is highly sensitive to the energy spectrum and particle distribution of individual SCR events. Webber et al.[2] have compiled new data sets of both integral and differential fluence spectra for major SCR events since 1940 and used these to calculate ^{10}Be production for each event.

Two significant SCR events occurred within the period covered by the DSS0506 snow pit record; these are 29 Oct 2003 and 20 Jan 2005. Integrated particle fluences and calculated polar production of ^{10}Be for these events are listed in Table 1. 29 Oct 2003 consisted of a series of events that were among the strongest in the current solar cycle (29 Oct 2003 events). The 20 Jan 2005 event exhibited high particle intensity in the higher energy region of the spectrum (i.e. a flatter spectrum) (20 Jan 2005 events). Both events caused ground-level enhancements (GLEs) in CRF recorded by neutron monitors, though the 29 Oct 2003 events had much smaller amplitudes. At peak intensity, the 20 Jan 2005 event resulted in a 2000% increase in CRF to the polar regions,[2] making it the strongest GLE in over 50 years. ^{10}Be production for the 29 Oct 2003 events peaked at $\sim$100 MeV, whereas the flatter spectrum of the 20 Jan 2005 event moved the production peak to higher energies.[2]

In terms of ^{10}Be production, SCRs during the 29 Oct 2003 event were calculated to contribute an additional $\sim$5% (with respect to GCR-produced

Table 1.   Calculated polar ^{10}Be production and particle fluences (F) for 29 Oct 2003 and 20 Jan 2005 SCR events.

Event	GLE peak intensity %[a]	Relative increase in annual ^{10}Be production[b]	Relative increase in monthly ^{10}Be production[c]	$F > 350$ MeV	$F > 100$ MeV	$F > 10$ MeV
29 Oct 2003	10, 35, 16	5%	60%	2.5×10^6	9.0×10^7	1.0×10^{10}
20 Jan 2005	2000	10%	120%	1.5×10^7	6.0×10^7	8.0×10^8

[a]Increases in peak intensity ("GLE" or Ground Level Enhancement) recorded by neutron monitors at sea level for the polar regions.
[b]The % SCR contribution to ^{10}Be production with respect to GCR-produced ^{10}Be. All % contributions are expressed relative to GCR production during 2003 (the year of the "reference event"; 29 Oct 2003, considered by Webber et al.[2]
[c]Relative monthly contributions are calculated by attributing all SCR production to the month of the SCR event. All % monthly contributions are expressed relative to mean monthly GCR production for 2003, assuming GCR production was evenly distributed over the year. Table adapted from Webber et al.[2]

^{10}Be) to polar production in 2003, or an additional ~60% to polar production during October.[2] The 20 Jan 2005 event was calculated to contribute an additional ~10% to polar production for 2005 or an additional ~120% of polar production for January.[2]

Here we look for a concentration signal in the snow pit record related to these two SCR events. This work is greatly facilitated by the theoretical predictions of SCR contribution to polar production of ^{10}Be as estimated by Webber *et al.*[2]

Figure 3(c) shows hourly mean CRF data[32] recorded at the McMurdo neutron Monitor. The GLEs associated with 29 Oct 2003 and 20 Jan 2005 events show up clearly as spikes in CRF intensity (note the greater amplitude of the 20 Jan 2005 event). The relationship between measured ^{10}Be concentration in the DSS0506 snow pit (Fig. 3(a)) and the timing and amplitude of modeled ^{10}Be production (Table 1) is considered for each event in turn.

3.3.1. *29 Oct 2003 SCR events*

^{10}Be concentrations do not show any significant deviation from the mean following the 29 Oct 2003 events (Fig. 3(a)). This may be partly due to the smaller amplitude of the production increase associated with the event (approximately half the amplitude of 20 Jan 2005 according to Webber *et al.*[2]). However, we expect the principal reason is that meteorological influences in the 1–2 months following the event are not conducive to rapid transport of a production signal to the ice core record.

3.3.2. *20 Jan 2005 SCR event*

A sharp peak in ^{10}Be concentration is observed following the 20 Jan 2005 event (Fig. 3(a)). ^{10}Be concentrations in the two samples following the event were 12.6 and 10.5×10^3 atoms g^{-1} (with 4% analytical errors); the mean for these samples is $>4\sigma$ above the mean for the remainder of the data set. The ^{10}Be peak was dated to mid-February 2005, indicating a delay of ~1 month between the SCR event and the maximum in ^{10}Be concentration. Our dating of the concentration peak is reliable and supported by meteorological data from nearby Casey station and prior work.[25,33] Significantly, there was precipitation throughout the summer period, confirming that the δ^{18}O record used in dating (refer to Sec. 2.4) was reliable for the period that included the event. Furthermore, the monthly mean temperature record for Casey confirms that the annual temperature maxima occurred in January

2005, consistent with the timing of the maximum in $\delta^{18}O$ assumed in dating (refer also to Sec. 4.4).

The length of delay between the production signal and concentration signal in the ice core record depends on atmospheric mixing and deposition processes. We note that a delay of $\sim$1 month is similar to the reported tropospheric residence time of ^{10}Be.[17] The approximate doubling in ^{10}Be concentration following the event is consistent with an approximate doubling in ^{10}Be production caused by the SCR event for January. Hence, the timing and amplitude of the ^{10}Be peak are consistent with a connection to the 20 Jan 2005 event. However, variable influences of meteorological effects on the record make it difficult to ascertain causality. We note that the peak occurs in summer when there is evidence that ^{10}Be arrival to polar sites is augmented by meteorological effects. Meteorological conditions at the time of the two SPE events are now discussed.

4. Discussion

4.1. *Atmospheric transport and deposition of SCR-produced ^{10}Be*

In the polar regions ^{10}Be production dominantly occurs in the stratosphere; for example, Masarik and Beer[30] calculate that stratospheric production of ^{10}Be over the poles may account for up to eight times tropospheric production. In addition, the energies and fluxes of energetic particles entering the atmosphere following large solar events are such that significant short-timescale (typically 10s of minutes to hours) enhancement of spalation products takes place preferentially in the lower stratosphere at polar latitudes.[2]

Following production, ^{10}Be is rapidly scavenged by aerosol particles.[17] In the stratosphere, the aerosol population is primarily comprised of sulphate particles, with a small admixture of meteoritic material.[34,35]

The concentration of ^{10}Be in ice core records involves interplay between production, transport, and deposition processes. We propose that observation of enhanced ^{10}Be concentrations in ice at Law Dome subsequent to SCR events (i.e. a SCR production signal) requires (a) that vertical transport permits polar stratospheric aerosols tagged with ^{10}Be to mix down into the troposphere before meridional mixing has damped out the production signal in the atmosphere, and (b) that precipitation processes are favorable for scavenging aerosols tagged with ^{10}Be and depositing them to the ice sheet. The amplitude of the SCR production signal that is sufficient to be observed in the ice core record will depend on the efficacy of these processes. We assess the two SCR events in the snow pit record with respect to these criteria using dispersion modeling of stratospheric aerosol pathways, empirical data on stratospheric aerosol arrival to the Antarctic ice sheet, and meteorological data on snow deposition in the Law Dome region (Secs. 4.2–4.4).

4.2. *Dispersion modeling of stratospheric aerosol pathways*

To assess the extent of meridional and vertical transport of the production signal for each SCR event, we used the dispersion component of the Hybrid Single-Particle Lagrangian Integrated Trajectory (HYSPLIT) model,[36] and performed forward analysis of simulated aerosol pathways. A population of approximately 2400 spherical particles of density $1.7\,\mathrm{g\,cm^{-3}}$ and $0.5\,\mu\mathrm{m}$ diameter (to simulate sulphate aerosols within the size range reported to predominantly scavenge ^{10}Be (McHargue and Damon[17]) were released from 41 sites spread evenly on a $1000\,\mathrm{km}$ grid south of $\sim 60°\,\mathrm{S}$. Particles were released over a period of 5 h approximating the duration of the peak solar particle fluxes for the 20 Jan 2005 and 29 Oct 2003 SCR events. Different release heights were trialed consistent with the expected height of ^{10}Be production.[2] The HYSPLIT model was used to trace the location of particles at 24 h intervals until subsidence to the troposphere occurred (subsidence to troposphere was judged with reference to NCEP/NCAR reanalysis data and using the World Meteorological Organisation thermal tropopause definition). Simulations spanned 90 days of simulated time.

Results from the dispersion modeling suggested significant aerosol subsidence to the troposphere within weeks to months after particle release for all simulations. This is rapid with respect to the 1- to 2-year mean stratospheric residence time of ^{10}Be given by Koch and Mann[37] and Dibb et al.[38] Results suggest slightly stronger subsidence for the 29 Oct 2003 events. However, a coherent signal was more likely to be transmitted for ^{10}Be produced in the lowermost part of the stratosphere (200 hPa level, $\sim 12.5\,\mathrm{km}$ altitude), and the higher intensity of high-energy particles associated with the 20 Jan 2005 event (Table 1) is expected to cause greater production in this region compared to the 29 Oct 2003 event.

The modeled aerosol descent would be expected to transmit a SCR production signal to the troposphere for both SCR events. Support for the relatively rapid aerosol descent interpreted from our model results is offered by a recent global circulation model experiment, which suggested incomplete meridional mixing of ^{10}Be produced over the poles.[15] A detailed consideration of modeled stratospheric aerosol dispersion following the SCR events will be presented in a separate manuscript (Klekociuk et al., in prep).

4.3. Empirical data on arrival of stratospheric aerosols to polar regions

An important difference between the meteorological conditions over Antarctica for the two events is the condition of the polar vortex. The polar vortex usually forms in late May and persists until between October and December. Atmospheric dynamics within the vortex are dependent on the altitude and latitude under consideration, and exhibit considerable interannual variability.[39] In general, while the vortex is present, temperature gradients across the tropopause are weak, and enhanced subsidence of stratospheric air is expected.[40] Consideration of vortex area data (for example, vortex area analyses for the 450 K potential temperature surface in the lower stratosphere[41]) shows that the 29 Oct 2003 SCR occurred while the vortex was breaking down and the 20 Jan 2005 event occurred well after the vortex of the 2004 winter had dissipated. One might expect that enhanced subsidence of stratospheric air while the vortex is present would promote the arrival of stratospheric aerosols to the polar regions. However, numerous empirical data suggest that stratospheric aerosols predominantly arrive to the Antarctic ice sheet in late summer. For example, ground-based measurements of the stratospheric aerosol markers ^{7}Be/^{10}Be ratio and ^{7}Be/^{210}Pb ratio[42] show late summer maximum at Neuymaer[29,43] as do ^{7}Be levels at South Pole and coastal Terre Adélie.[42,44,45] ^{10}Be concentrations in snow also show evidence of enhancement during late summer and autumn, for example at Law Dome[11] and Dye 3, South Greenland.[28] No mechanism has been clearly demonstrated to explain the summer to autumn maximum in stratospheric aerosol arrival to polar regions. Wagenbach[29] suggested that increased vertical mixing may occur at this time due to weakening of the tropospheric surface temperature inversion, whereas Beer et al.[28] proposed that a summer to autumn peak may be linked to longer-range transport of stratospheric aerosol from mid-latitude tropospheric injections.

Our aerosol dispersion results suggest that meridional mixing following the two SCR events was probably insufficient to damp out the production signals. Empirical data from aerosol and ice core measurements of ^{10}Be suggests conditions would be more favorable for the arrival of stratospheric aerosols to the Law Dome in summer following the 20 Jan 2005 SCR rather than the 29 Oct 2003 event. Deposition processes are now considered.

4.4. *Deposition processes*

Incorporation of [10]Be into polar ice sheets is by a combination of dry fallout of [10]Be containing aerosols ("dry" deposition) and scavenging of [10]Be containing aerosols by precipitation ("wet" deposition).[46] The relative contribution of wet to dry precipitation is site- and climate-specific.[15] Studies of [10]Be deposition at Law Dome have found that wet deposition, i.e. the arrival of aerosol particles with precipitation is dominant.[11,27] For this site, registration of a concentration signal in high-resolution [10]Be data will be more likely if significant aerosol scavenging by precipitation occurred in the period following the SCR event.

We consulted precipitation records from the Australian Casey Station, the nearest meteorological station to Law Dome (~100 km away). Prior inter-comparison of snow pit, ice core, and meteorological data from Law Dome and Casey station has demonstrated that meteorological observations at Casey can inform on meteorological conditions in the Law Dome region.[25,33] Precipitation in February 2005 following the 20 Jan 2005 event was high (22.8 mm), 65% above the 30-year mean, and 70% of the precipitation at Casey during the 3-month period spanning the event (December, January, February) arrived in February. By contrast, the precipitation in the month following the 29 Oct 2003 events was low (12 mm), slightly below the 30-year mean, and only 30% of the precipitation at Casey during the 3-month period spanning the events arrived in November. These data suggest that aerosol scavenging by precipitation is likely to have been facilitated in the period following the 20 Jan 2005 event and not so well facilitated following the weaker 29 Oct 2003 events.

5. Summary and Conclusions

Measured [10]Be concentrations in the 2.7-y DSS0506 snow pit were compared with CRF data from the McMurdo neutron monitor and with modeled [10]Be productions from SCR events. There was no relationship observed between measured [10]Be concentration and monthly mean CRF data. This result confirms that over short (sub-annual to annual) timescales, [10]Be concentration in ice core records is mainly influenced by meteorological effects. However, we find evidence that the sharp production signals associated with SCR events may be imprinted on ^{10}Be concentration.

We investigated the possible connection of ^{10}Be concentration and modeled the ^{10}Be production for the SCR events of 29 Oct 2003 and 20 Jan 2005. A strong peak in concentration was observed $\sim$1 month after the 20 Jan 2005 event but no increase in concentration was observed following the 29 Oct 2003 event. We argued that atmospheric conditions and deposition processes following the 20 Jan 2005 event were likely to be favorable for registration of the production signal, whereas conditions following the 29 Oct 2003 events were less favorable. Aerosol dispersion modeling suggested that for both events ^{10}Be aerosols tagged with ^{10}Be reached the troposphere before excessive smoothing of the production signal occurred. The 20 Jan 2005 event exhibited high fluence in the high-energy part of the spectrum; hence, production was expected to have occurred lower in the stratosphere from where mixing into the troposphere was more efficient. Furthermore, empirical data from ice core and aerosol measurements at Antarctic stations show that stratospheric aerosols are preferentially delivered to the polar region in the late summer, which corresponds with the observed February maximum in ^{10}Be concentration from the 20 Jan 2005 event. Finally, precipitation data from nearby Casey station shows that precipitation was particularly high in February following the 20 Jan 2005 event and suitable for leaching of the production signal from the atmosphere, as opposed to November following the 29 Oct 2003 SCR when precipitation was low. Considering these transport and deposition factors along with the greater amplitude of ^{10}Be production by the 20 Jan 2005 SCR, it seems reasonable that only the 20 Jan 2005 SCR was observed in ^{10}Be concentrations. It should be noted, however, that we cannot rule out that the strong peak in concentration during February 2005 has some other cause.

Identification of production signals in ^{10}Be data is complex due to the sensitivity of ^{10}Be concentration to variable climate and meteorological processes. A detailed analysis over a longer record is required to constrain the link between measured ^{10}Be concentrations and SCR production of ^{10}Be. We plan to carry out such an analysis using the most recent 60 years of the high-resolution DSS0506 ice core record.

Acknowledgments

This work was assisted by the Australian Government's Cooperative Research Centres Programme, through the Antarctic Ecosystems and

Climate Cooperative Research Centre (ACE CRC). The Australian Antarctic Division supported the field campaign to sample the DSS0506 snow pit and ice core records. The Cosmogenic Climate Archives of the Southern Hemisphere (CcASH) project of the Australian Nuclear Science and Technology Organisation (ANSTO) supported ^{10}Be measurements. Joel Pedro is grateful for research funding from an Australian Institute of Nuclear Science and Engineering Postgraduate Research Award and an Australian Postgraduate Award. The Bartol Research Institute neutron monitor program is supported by National Science Foundation grant ATM-0527878.

References

1. I. G. Usoskin, S. K. Solanki, G. A. Kovalstsov, J. Beer and B. Kromer, *Geophys. Res. Lett.* **33** (2006) L08107.
2. W. R. Webber, P. R. Higbie and K. G. McCracken, *J. Geophys. Res.* **112** (2007) A10106, doi: 10.1029/2007JA012499.
3. D. Lal and B. Peters, *Handbuch der Physik* **46** (1967) 551.
4. G. M. Raisbeck and F. Yiou, in *The Ancient Sun: Fossil Record in the Earth, Moon and Meteorites; Proceedings of the Conference*, eds. R. O. Peppin, J. A. Eddy and R. B. Merill (Pergamon Press, New York and Oxford, 1980), pp. 185–190.
5. R. Muscheler, F. Joos, J. Beer, S. A. Muller, M. Vonmoos and I. Snowball, *Quart. Sci. Rev.* **26** (2007) 82–97.
6. G. Wagner, J. Masarik, J. Beer, S. Baumgartner, D. Imboden, P. W. Kubik, H.-A. Synal and M. Suter, *Nucl. Instrum. Methods, Phys. Res. Sect. B* **172**(1–4) (2000) 597.
7. R. Muscheler, J. Beer, P. W. Kubik and H. A. Synal, *Quart. Sci. Rev.* **24**(16–17) (2005) 1849.
8. E. Bard and M. Frank, *Earth Planet. Sci. Lett.* **248** (2006) 1.
9. C. M. Ammann, F. Joos, D. S. Schimel, B. L. Otto-Bliesner and R. A. Tomas, *Proc. Nat. Acad. Sci. USA* **104**(10) (2007) 3713.
10. J. Beer, *Space Sci. Rev.* **94** (2000) 53.
11. J. Pedro, T. van Ommen, M. Curran, V. Morgan, A. Smith and A. McMorrow, *J. Geophys. Res.* **111**(D10) (2006) D21105, doi: 10.1029/2005JD006764.
12. E. J. Steig, P. J. Polissar, M. Stuiver, R. C. Finkel and P. M. Grootes, *Geophys. Res. Lett.* **25** (1996) 523.
13. A. Aldahan, G. Possnert, S. J. Johnsen, H. B. Clausen, E. Isaksson, W. Karlen and M. Hansson, *Proc. Ind. Acad. Sci. (Earth Planet. Sci. Lett.)* **107**(2) (1998) 139.
14. H. Moraal, R. Muscheler, L. du Piessis, P. W. Kubik, J. Beer, K. G. McCracken and F. B. McDonald, *S. Afr. J. Sci.* **101** (2005) 299.
15. C. V. Field, G. A. Schmidt, D. Koch and C. Salyk, *J. Geophys. Res.* **111** (2006) D15107, doi: 10.1029/2005JD006410.

16. K. G. McCracken, *J. Geophys. Res.* **109** (2004) A04101, doi: 10.1029/ 2003JA010060.
17. L. R. McHargue and P. E. Damon, *Rev. Geophys.* **29**(2) (1991) 141.
18. M. A. J. Curran, T. D. van Ommen and V. I. Morgan, *Ann. Glaciol.* **27** (1998) 385.
19. A. J. McMorrow, T. D. van Ommen, V. Morgan and M. A. J. Curran, *Ann. Glaciol.* **39** (2004) 34.
20. V. I. Morgan, C. W. Wookey, J. Li, T. D. van Ommen, W. Skinner and M. F. Fitzpatrick, *J. Glaciol.* **43** (1997) 3.
21. D. Child, G. Elliott, C. Misfud, A. M. Smith and D. Fink, *Nucl. Instrum. Methods Sect. B* **172** (2000) 856.
22. D. Fink, B. McKelvey, D. Hannan and D. Newsome, *Nucl. Instrum. Methods Sect. B* **172** (2000) 838.
23. D. Fink and A. Smith, *Nucl. Instrum. Methods Sect. B* **259** (2007) 600.
24. W. Dansgaard, *Tellus* **16** (1964) 436.
25. T. D. van Ommen and V. Morgan, *J. Geophys. Res.* **102**(D8) (1997) 9351.
26. A. J. McMorrow, M. A. J. Curran, T. D. van Ommen, V. Morgan and I. Allison, *Ann. Glaciol.* **35** (2002) 463.
27. A. M. Smith, D. Fink, D. Child, V. A. Levchenko, V. Morgan, M. Curran and D. Etheridge, *Nucl. Instrum Methods Sect. B* **172** (2000) 847.
28. J. Beer *et al.*, *Atmos. Environ.* **25A**(5/6) (1991) 899.
29. D. Wagenbach, in *Chemical Exchange Between the Atmosphere and Polar Snow*, eds. E. W. Wolff and R. C. Bales, NATO ASI Series, Vol. I 43 (Springer-Verlag, Berlin Heidelberg, 1996), pp 173–199.
30. J. Masarik and J. Beer, *J. Geophys. Res.* **104**(D10) (1999) 12099.
31. W. R. Webber and P. R. Higbie, *J. Geophys. Res.* **108**(A9) (2003) 1355, doi: 10.1029/2003JA009863.
32. Bartol Research Institute, Neutron Monitor Data for McMurdo, http:// neutronm.bartol.udel.edu/.
33. A. J. McMorrow, PhD thesis, University of Tasmania, Hobart, Australia, 2006.
34. R. C. Whitten, O. B. Toon and R. P. Turco, *Pageoph.* **118** (1980) 86.
35. D. M. Murphy, D. S. Thompson and M. J. Mahoney, *Science* **282** (1998) 1664.
36. R. R. Draxler, *J. Appl. Meteorol.* **42** (2003) 308.
37. D. M. Koch and M. E. Mann, *Tellus, Ser. B* **48** (1996) 387.
38. J. E. Dibb, D. L. Meeker, R. C. Finkel, J. R. Southon, M. W. Caffee and L. A. Barrie, *J. Geophys. Res.* **99**(12) (1994) 855.
39. G. L. Manney, R. W. Zurek, A. ONeill and R. Swinbank, *J. Atmos. Sci.* **51**(20) (1994) 2973.
40. W. Schwerdtfeger, in *Weather and Climate of the Antarctic* (Elsevier, New York, 1984), p. 261.
41. NOAA, National Weather Service, National Centre for Environmental Prediction, Climate Prediction Centre, Time Series of the Southern Hemisphere Polar Vortex at 450K, http://www.cpc.ncep.noaa.gov/products/ stratosphere/polar/polar.shtml.

42. G. M. Raisbeck, F. Yiou, M. Fruneau, J. M. Loiseaux, M. Lieuvin and J. C. Ravel, *Geophys. Res. Lett.* **8** (1981) 1015.

43. D. Wagenbach, U. Goerlach, K. Moser and K. O. Muennich, *Tellus, Ser. B* **40** (1988) 426.

44. W. Maenhaut, W. H. Zoller and D. G. Coles, *J. Geophys. Res.* **84** (1979) 3131.

45. J. Sanak, G. Lambert and B. Ardouin, *Tellus, Ser. B* **37** (1985) 109.

46. G. M. Raisbeck and F. Yiou, *Ann. Glacio.* **7** (1985) 138.

Advances in Geosciences
Vol. 14: Solar Terrestrial (2007)
Eds. Marc Duldig *et al.*
© World Scientific Publishing Company

ATMOSPHERIC CHEMISTRY EFFECTS
OF THE 20 JANUARY 2005 SOLAR PROTON EVENT

A. R. KLEKOCIUK*, D. J. BOMBARDIERI*,†, M. L. DULDIG*
and K. J. MICHAEL†

*IOAC Program, Australian Antarctic Division
203 Channel Highway, Kingston, Tasmania 7050, Australia

†Institute of Antarctic and Southern Ocean Studies
University of Tasmania, Hobart, Tasmania 7001, Australia

Atmospheric ionization during Solar Proton Events (SPE) has traditionally been modeled using top-of-atmosphere (TOA) particle fluxes measured by near-Earth spacecraft, and as a consequence the modeling has been restricted to particle energies below $\sim 500\,\mathrm{MeV}$. However, as inferred from measurements by Earth-based muon telescopes and neutron monitors during Ground Level Enhancement (GLE) events, protons accompanying the most energetic SPEs can have significant short-term ($< 1\,\mathrm{day}$ duration) fluxes at energies up to at least $20\,\mathrm{GeV}$. Here, we examine atmospheric chemistry changes in the polar atmosphere during the large GLE of 20 January 2005. We use TOA particle spectra derived from satellite measurements augmented with proton spectra inferred from ground-based neutron monitor data to drive NO_y and OH production in the SOCRATES two-dimensional atmospheric chemistry model. We show that the energetic particles of the January 20 event and associated particle enhancements during the preceding 5 days produced sufficient short timescale ionization in the lower stratosphere for measurable effects on HNO_3 and O_3 concentrations in the polar regions of both hemispheres. We also show that the detailed consideration of the most energetic particles that contributed to this GLE is of low importance for modeling ionization in the lower stratosphere, at least on a daily average timescale.

1. Introduction

Energetic particles originating outside the Earth's atmosphere play an important role in perturbing the chemistry and thermodynamics of the middle atmosphere.[1] The specific effects include the production of reactive chemical species which influence the concentration of radiatively-important gases such as ozone, as well as direct heating. The sources of the particles include galactic and solar cosmic rays, auroral and relativistic electrons, and energetic solar protons. Of these sources, the protons which impinge on the atmosphere during Solar Proton Events (SPEs) cause the most pronounced perturbations to the middle atmosphere.[1-5]

During an SPE, charged particles gain access to the Earth's magnetic field from the interplanetary magnetic field, and precipitate into the atmosphere, particularly in the polar regions (poleward of ~55 degrees geomagnetic latitude). Protons with energies above a few hundred MeV can penetrate into the lower stratosphere.[4] The precipitating particles interact with the neutral atmosphere to produce ionizations, dissociations, dissociative ionizations, and excitations. These interactions lead to the production of odd hydrogen ($HO_x = H + OH + HO_2$) and odd nitrogen ($NO_y = N + NO + NO_2 + NO_3 + HO_2NO_2 + HNO_3 + 2N_2O_5 + ClONO_2$) chemical species which catalytically destroy ozone[1,6-10] on timescales from hours to months or years.[11-14] These ozone losses, while they do not significantly perturb the solar ultraviolet irradiance at Earth's surface, have been linked with variations in temperatures and winds in the polar middle atmosphere.[1,15-19]

To date, the modeling of SPE effects on the atmosphere has relied exclusively on spacecraft measurements of particle fluxes in the near-Earth environment.[5,20,21] The spacecraft sensors provide proton flux measurements for energies up to ~500 MeV. This is normally adequate to cover the SPE proton energy range which is typically 10–100 MeV. However, as determined from observation of Ground-Level Enhancement (GLE) events in the cosmic ray flux measured by Earth-based neutron monitors and muon telescopes, certain SPEs can have significant fluxes of protons with energies greater than 1 GeV. Neutron monitors and muon telescopes measure the flux of secondary particles created in the atmosphere by the interaction of the SPE protons with the nuclei of atmospheric gases. An established technique is available to infer the energy spectrum of the incident protons during GLE events from the neutron and muon records.[22] Importantly, the SPEs accompanied by a GLE are the most energetic events and have the greatest potential for effects deep in the atmosphere.

Here we examine the atmospheric effects of energetic particles from one of the most significant solar events of the modern era, namely, the large solar particle event of 20 January 2005, and examine whether consideration of the GLE particles is of any importance for atmospheric chemistry modeling. The remainder of this chapter is structured as follows. In Sec. 2 we outline the circumstances of the 20 January 2005 solar event, and follow this in Sec. 3 with the description of our atmospheric ionization and chemistry modeling for this event, and analysis of comparative atmospheric trace gas measurements by the Aura satellite. In Sec. 4, we compare the modeling and observational results, and this is followed by our conclusions.

2. The 2005 January 20 Solar Event

During January 2005, National Oceanic and Atmospheric Administration (NOAA) solar active region 10720 produced a series of significant X-ray flares which resulted in episodic proton enhancements in the near-Earth environment spanning the period 15–21 January. The most significant of these flares, with X-ray/optical classification X7.1/2B, occurred on 20 January and produced the hardest and most significant SPE of solar cycle 23. The X-ray flare began at 06:36 UT and peaked at 07:01 UT. It was accompanied by a GLE, event number 69, which was detected by ground-based neutron monitors from 06:48 UT.[23] The peak of the GLE occurred within 5 min at many stations. The largest sea-level corrected neutron monitor response was observed at Terre Adélie (Antarctica) with a maximum enhancement of ∼4250% above the pre-event level. The relaxation phase was initially relatively rapid; for example, the decline to 10% of the peak value at Terre Adélie occurred within ∼70 min. There was subsequently a slower relaxation over several hours. For example, by 14:00 UT the enhancement above the galactic cosmic ray (GCR) background at Terre Adélie was still ∼10%.

There are two notable features of GLE 69. Firstly, the peak flux of relativistic solar particles was the highest observed since 1956.[24] Secondly, the response of the global neutron monitor network indicates that the pitch angle distribution of the relativistic particles was both highly anisotropic and hemispherically asymmetric, with notably larger fluxes (normalized to sea level) observed in the Antarctic sector between the neutron monitor sites of Terre Adélie, McMurdo, and South Pole.[23]

At geosynchronous satellite altitudes, ≥100 MeV protons measured by Geostationary Operational Environmental Satellites (GOES) peaked at 07:10 UT, and attained the highest flux level observed since October 1989.[25] The ≥ 100 MeV proton event ended at 18:45 UT on 21 January.

3. Ionization Modeling

We model the atmospheric ionization resulting from precipitating protons and alpha particles following the approach used by Vitt and Jackman[26] and Decker *et al.*[27] This approach uses the "continuous slowing down approximation" to degrade the incoming particle flux as it penetrates into the atmosphere. The degradation process was evaluated using particle stopping power data from Asher.[28]

3.1. *The particle spectrum*

Our primary source of data on the top-of-atmosphere (TOA) particle flux is inferred from measurements by the GOES-11 spacecraft. We used 5-min resolution differential flux measurement of protons (from channels P2–P7, corrected for secondary particle effects) and alphas (from channels A1–A6, uncorrected for secondary effects[a]). We did not use data from the P1 proton channel (which covers the energy range of 0.8–4 MeV), because this channel is influenced by trapped particles in the outer zone of the magnetosphere, and these particles do not normally precipitate into the atmosphere. GOES-11 provided the best available spacecraft particle measurements for the January 20 GLE; alternative measurements by GOES-10 and Interplanetary Monitoring Platform (IMP) 8 were either not available or exhibited significant saturation.

The GOES-11 proton and alpha data were each interpolated into 50 logarithmically spaced energy bins over the range 6.5–500 MeV and 7–500 MeV, respectively. From 06:50 UT to 08:00 UT on 20 January, the differential proton channels below ~ 30 MeV exhibited saturation; the fluxes in these channels were adjusted through interpolation across channels at the lower and upper ranges of the affected energy range.

During the interval 06:50 UT to 12:00 UT on 20 January, the technique employed by Lovell *et al.*[22] and Bombardieri *et al.*[23] was used to estimate 5-min differential solar proton fluxes over the energy range 500 MeV–10 GeV from neutron monitor measurements. The satellite and neutron monitor proton spectra were merged to contain 72 logarithmically spaced bins up to 10 GeV, and interpolation was used to create a smooth transition between the two types of data at 500 MeV. Additionally, the background GCR proton spectrum for high solar activity from Smart and Shea[29] was included. Similarly, the satellite alpha particle spectra were extended to 10 GeV in 72 logarithmically spaced bins through inclusion of the alpha GCR spectrum from Smart and Shea.[29] In extending the energy range of the

---

[a]The uncorrected fluxes generally overestimate the true alpha flux. Because the alpha flux contributes less than 10% to the total ionization below ~ 50 km, uncertainities in the alpha flux are not considered significant for ionization in the lower stratosphere.

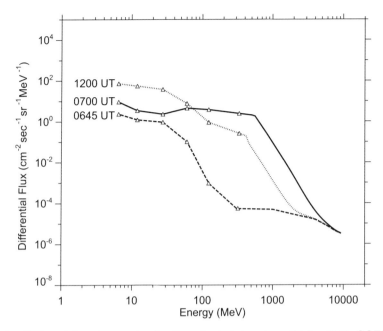

Fig. 1.   Differential proton spectra for three 5-min intervals on 20 Jan 2005. GOES-11 measurements are indicated by symbols. As described in the text, flux values at energies above 500 MeV have been estimated using background GCR flux data from Smart and Shea,[29] and additionally in the case of 07:00 UT and 12:00 UT, from neutron monitor measurements.

satellite measurements using the GCR background spectra, interpolation was used between 500 MeV and 3 GeV to provide a smooth transition. Examples of the synthesized proton spectra for selected time intervals are presented in Fig. 1.

3.2. Ion pair production

The particle pitch angle distribution was divided into 49 discrete angles. As mentioned above, the GLE particles showed evidence of an anisotropic pitch angle distribution for this event. In our modeling we assumed an isotropic pitch angle distribution, and will consider the anisotropy in detail in a future publication.

We evaluated the energy deposited in atmospheric slabs of 2 km vertical thickness. The background atmosphere was obtained from the MSISE90 climatology.[30] We used the mean density profile at 75° latitude and

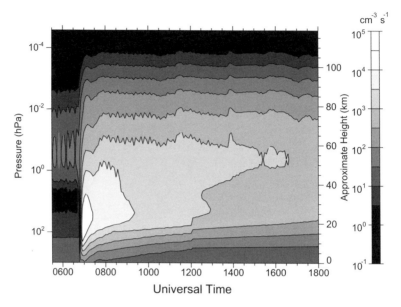

Fig. 2. Ion pair production rate evaluated for 20 Jan 2005 at latitude 75° as a function of time and pressure using 5-min inferred proton and alpha particle spectra.

assumed that the full particle spectrum would be able to gain access to this latitude. An ion pair creation threshold of 36 eV was assumed[31] and secondary particle production (e.g. induced by X-ray emission from electron bremsstrahlung as discussed by Schröter et al.[32]) was not considered. A time–height diagram of modeled ion pair production rate (IPPR) is presented in Fig. 2, in which a rapid change in ionization is apparent in the lower stratosphere (below ∼30 km height) prior to 08:00 UT.

Note that the low energy cutoff of our particle spectra causes underestimation of the ionization above heights of approximately 75 km, while the upper energy cutoff of 10 GeV is not likely to be of any consequence to the overall ionization profile owing to the low flux levels at these energies, even during the GLE.

3.3. Atmospheric chemistry modeling

We used version 3 of the SOCRATES two-dimensional atmospheric chemistry model[33,34] to examine the atmospheric ionization effects on trace gas concentrations. We modified the model to use our daily average IPPR profiles as a driver of NO_y and HO_x reactions. The model was run with mainly default parameters, which included default initialization of the background chemical species. However, we included Polar Stratospheric Cloud heterogeneous reactions (or significance for the Arctic atmosphere at that time of the year), and set the archival time-step for model results to 1 day. The latitudinal variation in IPPR due to the effects of the geomagnetic field was parameterized by scaling the IPPR calculated for 75° latitude using a rigidity scaling from Heaps.[35] We assumed scaling factors converting IPPR to production rates of OH and NO_y as 1.9 and 1.25, respectively.[9]

In Fig. 3 we show selected daily profiles of the estimated NO_y production to highlight the significance of 20 January in comparison to the effects on other days when solar particle fluxes were elevated, and also the significance of considering the high energy end of the proton spectrum. At 25 km height, the daily NO_y production on 20 January was estimated to be 300 times the energetic particle pre-event background level. The significance of the GLE event can be gauged by comparing the January profile with that labeled "a" for which the high energy proton spectrum only includes the estimated GCR background. The ratio of profile "a" to that of 20 January is 0.98 at 25 km and 0.35 at 15 km, implying that detailed consideration of the GLE particles for this event, at least on daily resolution, is of little consequence above 25 km height. At lower altitudes, the significance of the GLE modeling becomes greater, but this is the region where gas-phase and

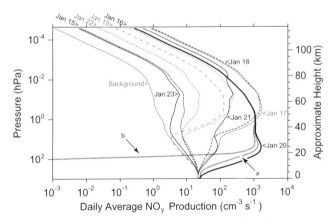

Fig. 3. Estimated daily average NO_y production rate at latitude 75° as a function of pressure for 15–23 January 2005. The background profile is the production rate averaged over period 1–14 January. Labels "a" and "b" refer to profiles for 20 January calculated without inclusion of neutron monitor data, and particle energies greater than 500 MeV, respectively.

heterogeneous reactions and transport strongly influence background trace gas concentrations. Note also that the ratio of the profile labeled "b" in Fig. 3 to that of 20 January is 0.71 at 25 km, indicating that even ignoring particles above 500 MeV (e.g. as in the analysis performed by Vitt and Jackman[26]) can be justified when modeling ionization in and above the lower stratosphere.

We performed two separate runs of the SOCRATES model for the period 1 January to 14 February: a "background" run where the daily vertical profile of the IPPR due to energetic particles was held constant at the average estimated for 1–14 January 2005, and an "evaluation" run where the IPPR was evaluated on a daily basis from the particle flux data. Time–height differences between the evaluation and background runs for the northern and southern polar regions are shown in the top panels of Figs. 4 and 5 for HNO_3 and O_3, respectively. These constituents were selected because they are directly influenced by HO_x and NO_y reactions, and global height-resolved measurements of their distribution are available from the Aura atmospheric chemistry satellite.

3.4. Atmospheric chemistry measurements

We compared the model results with measurements of HNO_3 and O_3 from the Microwave Limb Sounder (MLS) experiment on the Aura satellite. We used version 2 MLS data: for discussions on the validation of these measurements, see Santee et al.[36] and Livesey et al.[37] We produced zonal means on each measured pressure level over the latitude ranges $60°$–$85°$ and $35°$–$60°$, taking into account the data quality flags and caveats. Because HNO_3 is rapidly photolyzed by sunlight, we formed separate "night" and "day" averages, corresponding to solar zenith angles at the measurement location of greater than $95°$ and less than $85°$, respectively. We considered the period 1 January–15 February in each of the available measurement years 2005–2007, and used the measurements in 2006 and 2007 to create an average "climatology" to compare against 2005. Selected time–height analyses of the zonal averages are shown for HNO_3 and O_3 in the lower panels of Figs. 4 and 5, respectively.

4. Discussion

In Fig. 4(a), the modeling shows a strong enhancement in HNO_3 near the stratopause (near 55 km height or 0.2 hPa) in the Northern Hemisphere

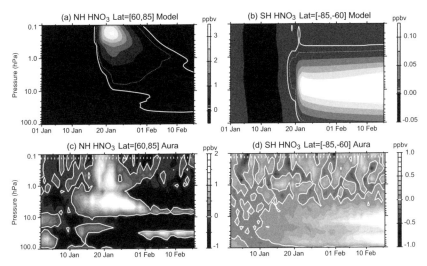

Fig. 4. Comparison of HNO₃ concentration changes derived from the model (panels (a) and (b) for Northern Hemisphere (NH) and Southern Hemisphere (SH), respectively) and Aura Microwave Limb Sounder (MLS) measurements (panels (c) and (d) for NH and SH, respectively). The model results represent the difference between the SPE simulation (using daily particle flux data) and a "background" simulation (where the particle fluxes were held constant at the average for 1–14 January) as described in the text. The Aura measurements represent the difference between daily averages (for "night" (NH) and "day" (SH)) and the average for 1–14 January. The white contours are levels of (a) 0.1 (0.2) ppbv for solid (dashed) lines, (b) 0.005 (0.01) ppbv for solid (dashed) lines, (c) 0 ppbv, and (d) 0 ppbv. The white vertical tick marks near the upper horizontal axis in (c) and (d) mark the mean times of the Aura measurements.

(NH) that commenced on 16 January (coincident with the most significant rise in the solar particle flux) and peaked on 22 January. Much of the polar cap was in darkness at this time, allowing HNO₃ to be relatively long-lived in the upper stratosphere. In Fig. 4(b), a much smaller enhancement is shown as occurring in the Southern Hemisphere (SH), which persists in the lower stratosphere where the lifetime of HNO₃ is longer compared to higher in the atmosphere.

The Aura measurements for the NH, shown in Fig. 4(c) reveal an obvious increase in the upper stratosphere commencing on 16 January, which is slightly smaller in magnitude, and at a similar height to the model result shown in Fig. 4(a). Two vertical enhancements on approximately 18 and 21 January are apparent, with the latter event being the strongest, which is consistent with the appearance of the model contours in Fig. 4(a). Note that the MLS HNO₃ measurements are generally of adequate

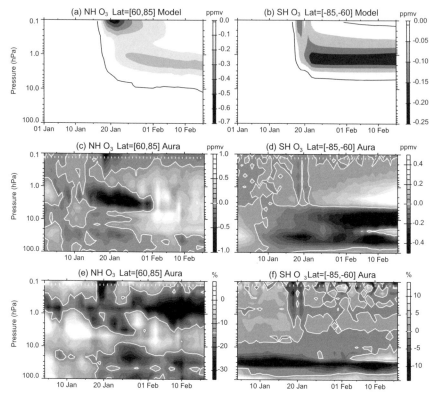

Fig. 5.   Similar to Fig. 4, but for O$_3$ concentration changes derived from the model (panels (a) and (b) for NH and SH, respectively) and Aura MLS measurements (panels (c), (e), and (d), (f) for NH and SH, respectively). For panels (e) and (f), the percentage change of the 2005 Aura measurements is shown relative to the daily mean averaged for 2006 and 2007 (where measurements exist on the same day for all years). In (a) and (b), the black contours are at −0.01 ppmv. In panels (c)–(f), the white contours are levels of (c) 0 ppmv, (d) 0 ppmv, (e) 0%, and (f) 0%.

signal-to-noise ratio only below 1 hPa,[36] and the SPE-related features above this level in Fig. 4(c) are only visible because of their intensity.

The strongest feature of Fig. 4(c), however, is a slowly descending enhancement that is apparent between 10 and 1 hPa. This feature becomes most apparent on 10 January (i.e. preceding the initial solar particle event) and reaches a maximum (relative to the 1–14 January mean) near 20 January. Examination of the climatology derived by averaging the corresponding period in 2006 and 2007, and 2005 data averaged over the eastern and western hemispheres (not shown) indicates that most of this feature may be associated with the upper levels of the unusually strong polar vortex that occurred during the 2004/05 boreal winter.[38] A secondary maximum in HNO_3 mixing ratio near 35 km height ($\sim$ 5 hPa), in addition to the primary maximum near 20 km height ($\sim$ 50 hPa), is a known feature of the polar vortex[39] that may be related to ion cluster chemistry associated with stratospheric aerosol. We suggest that the feature in Fig. 4(c) represents variability in the secondary HNO_3 maximum, and that the high energy solar particles that reach the lower stratosphere following the January 2005 solar events play a direct role in enhancing the layer.

The other features at heights below the 10 hPa pressure level in Fig. 4(c) likely relate to different compositional characteristics of the air masses inside and outside the polar vortex, and the dynamical variability in the location of the vortex with respect to the polar cap. These variations make very small expected HNO_3 enhancements in the lower stratosphere (indicated in Fig. 4(a)) problematic to detect. An improvement in our analysis would be to average each vertical level only within the confines of the polar vortex (as done in Orsolini *et al.*[39]). Note also that HNO_3 changes also were evident in the latitude band 60°N to 35°N latitude (not shown). Overall, the latitudinal variation in the HNO_3 changes may provide information on the degradation of the particle energies due to geomagnetic cutoff (in a similar manner to that examined in Verronen *et al.*[40]).

In the SH, the very weak enhancement suggested near 10 hPa in the model result of Fig. 4(b) is at or below the level of the day-to-day variability of the averaged MLS data shown in Fig. 4(d). Note in this figure the seasonal change in the lower stratosphere as the sun moves lower in the sky, and in particular, the two separate layers centered near 20 hPa and 60 hPa. There is however the suggestion of an abrupt change at the 0.1 ppbv level, comparable with the model results, near 20 January that reaches down in height to approximately 20 hPa ($\sim$25 km). It would be interesting to examine this feature in relation to NO_x levels that are available from other satellite sensors.[41] A change in HNO_3 or NO_x at this height that can be directly attributed to solar particles would provide a useful test of ionization modeling, and in particular the importance of secondary ionization effects from the highest energy particles.[42]

For O_3, the model results shown for the NH in Fig. 5(a) indicate a significant decrease in the upper stratosphere and mesosphere. Our model results are comparable with those of Verronen *et al.*[43] and indicate a measurable lower height limit of the SPE-related changes at the $\sim$5 hPa level ($\sim$35 km). For the NH, the SPE-related ozone changes are apparent in both Fig. 5(c) (which is referenced to the period 1–14 January) and 5(e) (which shows the percentage change from the 2006–2007 climatology). However, a large change is evidently also associated with the secondary HNO_3 layer shown in Fig. 4(c) and discussed above. Below the 10 hPa level, the ozone changes are most certainly related to enhanced depletion in the Arctic polar vortex driven by heterogeneous reactions.

In the SH, the O_3 changes are more abrupt and short-lived compared with the NH. The observed changes are at the 0.1 ppmv level, and are comparable with the model results. There is evidence of stronger changes in the eastern hemisphere compared with the western hemisphere (not shown) for the 20 January event, along the lines of the asymmetry in particle precipitation discussed by von Savigny et al.[44] A further line of investigation is to use the global neutron monitor network to elucidate the details of the asymmetry and its implications for longitudinal variation of atmospheric chemistry effects.

5. Conclusions

We have used an atmospheric ionization model for precipitating energetic particles to drive a relatively simple atmospheric chemistry simulation, and find reasonable agreement between our modeling and observed changes to HNO_3 and O_3 concentrations in the polar stratosphere of both hemispheres for the January 2005 solar particle events.

Measurable changes in HNO_3 and O_3 were evident down to the height of the ~5 hPa level, and there is a suggestion of changes down to the 20 hPa level over the southern polar cap in HNO_3 associated with the 20 January 2005 GLE. An enhancement in HNO_3 and a simultaneous decrease in O_3 were evident in a layer near 5 hPa in the northern polar cap that may relate to a feature of the polar vortex, namely, the secondary HNO_3 maximum, which was enhanced by atmospheric ionization from energetic solar particles.

We also find that detailed consideration of the flux of the highest energy particles accompanying the 20 January 2005 GLE indicates that these particles do not contribute in a significant manner to the total ionization in the lower stratosphere on the timescale of 1 day, owing to the short duration of the GLE event and in spite of its high peak flux. Despite this later finding, we have identified opportunities to further utilize our modeling of the solar particle fluxes at the highest energies to examine anisotropy in particle precipitation, and atmospheric effects on sub-day timescales.

Acknowledgments

The authors thank their colleagues at IZMIRAN (Russia) and The Polar Geophysical Institute (Russia) for contributing neutron monitor data. D. J. Bombardieri acknowledges receipt of an Australian Postgraduate Award and Australian Antarctic Science Scholarship. Neutron monitors of the Bartol Research Institute were supported by NSF ATM-0527878 and OPP-9724293 neutron monitor programs. GOES-11 data were obtained from Space Physics Interactive Data Resource http://spidr.ngdc.noaa.gov, and Aura/MLS data were obtained through the MLS web site at http://mls.jpl.nasa.gov. Part of this work was performed under Project 737 of the Australian Antarctic program.

References

1. C. H. Jackman and R. D. McPeters, in *Solar Variability and Its Effects on Climate*, Geophys. Monogr. Ser., Vol. 141, eds. J. M. Rapp and P. Fox (AGU, Washington, DC, 2004), p. 305.
2. C. E. Randall, D. E. Siskind and R. M. Bevilacqua, *Geophys. Res. Lett.* **28** (2001) 2385.
3. M. Sinnhuber, J. P. Burrows, M. P. Chipperfield, C. H. Jackman, M.-B. Kallenrode, K. F. Künzi and M. Quack, *Geophys. Res. Lett.* **30** (2003) 1818, doi: 10/1029/2003GL017265.
4. C. H. Jackman, M. T. DeLand, G. J. Labow, E. L. Fleming, D. K. Weisenstein, M. K. W. Ko, M. Sinnhuber and J. M. Russell, *J. Geophys. Res.* **110** (2005) A09S27, doi: 10.1029/2004JA010888.
5. G. Rohen, C. von Savigny, M. Sinnhuber, E. J. Llewellyn, J. W. Kaiser, C. H. Jackman, M.-B. Kallenrode, J. Schröter, K.-U. Eichmann, H. Bovensmann and J. P. Burrows, *J. Geophys. Res.* **110** (2005) A09S39, doi: 10.1029/2004JA010984.
6. L. H. Weeks, R. S. Cuikay and J. R. Corbin, *J. Atmos. Sci.* **29** (1972) 1138.
7. D. F. Heath, A. J. Krueger and P. J. Crutzen, *Science* **197** (1977) 886.
8. S. Solomon, G. C. Reid, D. W. Rusch and R. J. Thomas, *Geophys. Res. Lett.* **10** (1983) 257.
9. C. H. Jackman and R. D. McPeters, *J. Geophys. Res.* **90** (1985) 7955.
10. C. H. Jackman, A. R. Douglass, R. B. Rood and R. D. McPeters, *J. Geophys. Res.* **95** (1990) 7417.
11. W. Swider and T. J. Keneshea, *Planet. Space Sci.* **21** (1973) 1969.
12. P. A. Crutzen, I. S. A. Isaksen and G. C. Reid, *Science* **189** (1975) 457.
13. C. H. Jackman, J. E. Frederick and R. S. Stolarski, *J. Geophys. Res.* **85** (1980) 7495.
14. S. Solomon, D. W. Rusch, J.-C. Gerard, G. C. Reid and P. J. Crutzen, *Planet. Space Sci.* **29** (1981) 885.
15. P. M. Banks, *J. Geophys. Res.* **84** (1979) 6709.

16. G. C. Reid, S. Solomon and R. R. Garcia, *Geophys. Res. Lett.* **18** (1991) 1019.
17. J. Rottger, *COSPAR Colloq. Ser.* **5** (1992) 473.
18. R. M. Johnson and J. G. Luhmann, *J. Atmos. Terr. Phys.* **55** (1993) 1203.
19. C. H. Jackman, M. C. Cerniglia, J. E. Nielsen, D. J. Allen, J. M. Zawodny, R. D. McPeters, A. R. Douglass, J. E. Rosenfield and R. B. Rood, *J. Geophys. Res.* **100** (1995) 11641.
20. C. H. Jackman, R. D. McPeters, G. J. Labow, E. L. Fleming, C. J. Praderas and J. M. Russell, *Geophys. Res. Lett.* **28** (2001) 2883.
21. M. López-Puertas, B. Funke, S. Gil-López, T. von Clarmann, G. P. Stiller, M. Höpfner, S. Kellmann, G. M. Tsidu, H. Fisher and C. H. Jackman, *J. Geophys. Res.* **110** (2005) A09S44, doi: 10.1029/2005JA011051.
22. J. L. Lovell, M. L. Duldig and J. E. Humble, *J. Geophys. Res.* **103** (1998) 23733.
23. D. J. Bombardieri, M. L. Duldig, J. E. Humble and K. J. Michael, *Ap. J.* **682** (2008) 1315.
24. J. Bieber, J. Clem, P. Evenson, R. Pyle, M. Duldig, J. Humble, D. Ruffolo, M. Rujiwarodom and A. Saiz, *Proc. 29th Int. Cosmic Ray Conf.* (Pune) **1** (2005) 237.
25. National Oceanic and Atmospheric Administration (2008), http://umbra. nascom.nasa.gov/SEP.
26. F. M. Vitt and C. H. Jackman, *J. Geophys. Res.* **101** (1996) 6729.
27. D. T. Decker, B. V. Kozelov, B. Basu, J. R. Jasperse and V. E. Ivanov, *J. Geophys. Res.* **101** (1996) 26947.
28. J. Asher, in *Tables of Physical and Chemical Constants* (online edition), http://www.kayelaby.npl.co.uk/atomic_and_nuclear_physics/4_5/4_5_1.htm (2004), p. 1.
29. D. F. Smart and M. A. Shea, in *Handbook of Geophysics and the Space Environment*, ed. A. S. Jursa (Air Force Geophysics Laboratory, Cambridge Massachusetts, 1985), pp. 6-1–6-29.
30. A. E. Hedin, *J. Geophys. Res.* **96** (1991) 1159.
31. H. S. Porter, C. H. Jackman and A. E. S. Green, *J. Chem. Phys.* **65** (1976) 154.
32. J. Schröter, B. Heber, F. Steinhilber and M. B. Kallenrode, *Adv. Space Res.* **37** (2006) 1597.
33. G. Brasseur, M. H. Hitchman, S. Walters, M. Dymek, E. Falise and M. Pirre, *J. Geophys. Res.* **95** (1990) 5639.
34. A. K. Smith, *J. Atmos. Solar-Terr. Phys.* **66** (2004) 839.
35. M. G. Heaps, *Planet. Space Sci.* **26** (1978) 513.
36. M. L. Santee, A. Lambert, W. G. Read, N. J. Livesey, R. E. Cofield, D. T. Cuddy, W. H. Daffer, B. J. Drouin, L. Froidevaux, R. A. Fuller, R. F. Jarnot, B. W. Knosp, G. L. Manney, V. S. Perun, W. V. Snyder, P. C. Stek, R. P. Thurstans, P. A. Wagner, J. W. Waters, G. Muscari, R. L. deZafra, J. E. Dibb, D. W. Fahey, P. J. Popp, T. P. Marcy, K. W. Jucks, G. C. Toon, R. A. Stachnik, P. F. Bernath, C. D. Boone, K. A. Walker, J. Urban and D. Murtagh, *J. Geophys. Res.* **112** (2007) D24S40, doi: 10.1029/2007JD008721.

37. N. J. Livesey, M. J. Filipiak, L. Froidevaux, W. G. Read, A. Lambert, M. L. Santee, J. H. Jiang, H. C. Pumphrey, J. W. Waters, R. E. Cofield, D. T. Cuddy, W. H. Daffer, B. J. Drouin, R. A. Fuller, R. F. Jarnot, Y. B. Jiang, B. W. Knosp, Q. B. Li, V. S. Perun, M. J. Schwartz, W. V. Snyder, P. C. Stek, R. P. Thurstans, P. A. Wagner, M. Avery, E. V. Browell, J.-P. Cammas, L. E. Christensen, G. S. Diskin, R.-S. Gao, H.-J. Jost, M. Loewenstein, J. D. Lopez, P. Nedelec, G. B. Osterman, G. W. Sachse and C. R. Webster, *J. Geophys. Res.* **113** (2008) D15S02, doi: 10.1029/2007JD008805.

38. National Oceanic and Atmospheric Administration, http://www.cpc.ncep. gov/products/stratosphere/winter_bulletins/nh_04-05 (2005).

39. Y. J. Orsolini, G. L. Manney, M. L. Santee and C. E. Randall, *Geophys. Res. Lett.* **32** (2005) L12S01, doi: 10.1029/2004GL021588.

40. P. T. Verronen, C. J. Rodger, M. A. Clilverd, H. M. Pickett and E. Turunen, *Ann. Geophys.* **25** (2007) 2203, SRef-ID: 1432-0576/angeo/2007-25-2203.

41. M. López-Puertas, B. Funke, S. Gil-López, T. von Clarmann, G. P. Stiller, M. Höpfner, S. Kellmann, H. Fischer and C. H. Jackman, *J. Geophys. Res.* **110** (2005) A09S43, doi: 10.1029/2005JA011050.

42. P. T. Verronen, E. Turunen, Th. Ulich and E. Kyrölä, *Ann. Geophys.* **20** (2002) 1967.

43. P. T. Verronen, A. Seppälä, E. Kyrölä, J. Tamminen, H. M. Pickett and E. Turunen, *Geophys. Res. Lett.* **33** (2006) L24811, doi: 10.1029/2006GL028115.

44. C. von Savigny, M. Sinnhuber, H. Bovensmann, J. P. Burrows, M.-B. Kallenrode and M. Schwartz, *Geophys. Res. Lett.* **34** (2007) L02805, doi: 10.1029/2006GL028106.